BrightRED Study Guide

CfE HIGHER

MATHEMATICS

Linda Moon and Peter Richmond

First published in 2015 by:
Bright Red Publishing Ltd
1 Torphichen Street
Edinburgh
EH3 8HX

MIX
Paper from
responsible sources

FSC
www.fsc.org

FSC® C013254

A CIP record for this book is available from the British Library.

ISBN 978-1-906736-65-1

With thanks to:
PDQ Digital Media Solutions Ltd, Bungay (layout); Project One Publishing Solutions Ltd (copy-edit).
Cover design and series book design by Caleb Rutherford – e i d e t i c.

Acknowledgements
Every effort has been made to seek all copyright-holders. If any have been overlooked, then
Bright Red Publishing will be delighted to make the necessary arrangements.

Permission has been sought from all relevant copyright holders and Bright Red Publishing are
grateful for the use of the following:

Lagui/iStock.com (p 10); gordonsaunders/iStock.com (p 13); PinkBadger/iStock.com (p 13);
Images licensed by Ingram Image (pp 19, 29, 39, 94 & 95); albln/iStock.com (p 29);
OverKit/iStock.com (p 35); WesAbrams/iStock.com (p 65); Paul Martlew (p 81);
Caleb Rutherford e i d e t i c (pp 79, 87, 43 & 85); ssguy/Shutterstock.com (p93);
Exam questions taken from Higher Maths exam papers © Scottish Qualifications Authority
(n.b. solutions do not emanate from the SQA).

Printed and bound in the UK by Martins the Printers.

CONTENTS

INTRODUCTION

INTRODUCING CfE HIGHER MATHEMATICS

THE STRUCTURE AND AIM OF THIS BOOK

There is no short-cut to passing any course at Higher level. To obtain a good pass requires consistent, regular revision over the duration of the course. The aim of this book is to bring together, with the aid of examples, concise coverage of the course material. The book should be used in conjunction with your course notes and the knowledge gained from your classes.

In addition to the course content in unit order, there is a summary skills list provided at the end of the book. This list highlights the requirements for both unit and added value levels. Use this list to check your progress. You could use symbols, as shown in the table.

an empty box	I have not learned this
• a dot	we have covered this in class
– a dash	I understand this topic, but more practice is required
+ a cross	I have revised this area and I am confident with it.

The book uses these features:

Don't Forget boxes flag up vital pieces of information that you need to remember and important things that you must be able to do, plus some helpful hints.

Things to Do and Think About sections contain practice questions to test your understanding. The solutions to these tasks are available on the *Bright Red Digital Zone* (www.brightredbooks.net).

Online references and tests direct you to the *Bright Red Digital Zone*, an online source of examples and solutions covering the entire syllabus. The examples range in length and complexity, and include questions in the same style and format as you will meet in your exams. Solutions will be provided to all examples. This is a useful resource as you study throughout the session.

Websites which can be used to extend your knowledge. In addition, www.hsn.uk.net has summary notes, and the SQA website (www.sqa.org.uk) has information on the examination, past papers and solutions which may be useful.

Remember to ask your teacher for more advice if you get stuck. Your teacher will be pleased to see that you are serious about independent study and that you are taking responsibility for your learning. If your school offers homework clubs, clinics, after-school study or you have a tutor, keep a note of problems you have when working on your own and take the list or set of post-notes with you to get focused, expert help.

PRIOR KNOWLEDGE

As you prepare for Higher mathematics, you need to be aware that success in this hierarchical subject depends on your knowing many important topics from lower levels such as National 5.

You will benefit from regular practice in algebraic manipulation and from consolidating numerical skills. Manipulating surds and indices, negative numbers, basic fractions and the four rules of arithmetic are frequently required in questions. Many skills such as straight lines, vectors, completing the square and the discriminant will be revisited and extended.

You need to use correct notation and you should try to become familiar with mathematical terminology and vocabulary. The correct use of brackets is very important, particularly in cases where the meaning can be altered. Poor use can lead to an incorrect solution or could make reaching a solution impossible.

ONLINE

For a quick recap on topics you have forgotten, head to www.brightredbooks.net/N5Maths

ASSESSMENT

The Higher Mathematics course has unit assessments and an external Course assessment.

The Course assessment is a written examination consisting of two papers.

- Paper 1 (Non-calculator) lasts 70 minutes and has 60 marks.
- Paper 2 (Calculator) lasts 90 minutes and has 70 marks.

Both papers consist of short and extended response questions. Paper 1 usually has more questions of a shorter nature than Paper 2. The course is graded A (bands 1 and 2), B (bands 3 and 4), C (bands 5 and 6) or D (band 7) based on how well you do in the external examination.

Exam hints

You do not need to answer the questions in order. It may be better to choose a question that you can answer easily first, so that you settle your nerves. However, remember that the questions should be in order of difficulty. You should look for connections between parts of questions. These are almost always linked and, sometimes, an earlier result in part (a) or (b) is needed and its use avoids repeating work.

Communication is important in questions where standard results and formulae are used. It is insufficient simply to quote a result or formula: they need to be connected to the particular question. For example, using the limit in recurrence relations, it is not sufficient to simply quote $L = \frac{b}{1-a}$; it should be used in the context of the question.

You will not be told in every question to 'show your working', but you need to remember to be accurate, to give detail and to illustrate your understanding in your working. However, you should simplify expressions and try to use concise and efficient methods where possible.

Remember, the examination will contain unseen and unfamiliar questions and contexts. Do not let this put you off. It is important to get practice in carefully reading and interpreting problems so that you are able to apply your knowledge.

REVISION TIPS

General advice

- **Don't leave your revision until the last minute**. When you are still learning new topics, revise the ones you have already covered.
- Study for periods of between 30 and 45 minutes, unless you are doing a complete paper.
- **Take short breaks**, away from your study area, to keep your level of concentration high.
- During your study leave, build treats and relaxation time into your revision timetable. This will help you to focus and help you stick to your plan.
- In the run up to the exams, **Eat Well**, **Exercise Well** and **Sleep Well**.

Maths-specific revision tips

- The best way to revise mathematics is by doing it. There is a time for learning the necessary formulae and rules, but there is no substitute for practice.
- Once you have learned a topic or skill, try questions. Start off with straightforward questions, then Unit level, and progress to examination style. Test your knowledge on topic-based questions, then progress to a mixture of past-paper questions. It is important to recognise what to use and when, which skill to apply and where.
- Use the space in the margin of this book to add your own revision reminders.
- Mathematics is a subject to be practised often. If you complete one extra question every night in addition to your normal homework, you will reap the rewards. You will be able to ask for help the next day when the problem is fresh in your mind, and so you will quickly build up your knowledge and confidence.
- Mathematics also demands perseverance and time management – you will need to tackle a number of questions or a whole examination paper in one sitting.
- Mathematics is different from other subjects in so many ways – the good thing about revising it is that you can be active.

The best way to revise maths is to actively do it.

COURSE CONTENTS

All this content will be subject to sampling in the final examination.

Key ◆ US: Unit standards are skills that will be assessed in order to gain unit passes as you work through the course.	
➤ AV: Added Value material will not be assessed at unit level.	
Pink: Found in Algebra chapter (pp 8–35)	Green: Found in Geometry chapter (pp 58–73)
Yellow: Found in Calculus chapter (pp 36–57)	Blue: Found in Trigonometry chapter (pp 74–91)

UNIT 1: EXPRESSIONS AND FORMULAE

Applying algebraic skills to logarithms and exponentials: Manipulating algebraic expressions

◆ Simplifying a numerical expression, using the laws of logarithms and exponents

◆ Solving logarithmic and exponential equations

◆ Using the laws of logarithms and exponents

➤ Using a straight line graph to confirm relationships of the form $y = ax^b$, $y = ab^x$

➤ Solving for a and b equations of the following forms, given two pairs of corresponding values of x and y: $\log y = b\log x + \log a$, $y = ax^b$ and, $\log y = x\log b + \log a$, $y = ab^x$

➤ Modelling mathematically situations involving the logarithmic or exponential function

Applying trigonometric skills to manipulating expressions: Manipulating trigonometric expressions

◆ The addition or double angle formulae

◆ Trigonometric identities

◆ Converting $a\cos x + b\sin x$ to $k\cos(x \pm \alpha)$ or $k\sin(x \pm \alpha)$, $k > 0$, α in the first quadrant

➤ Converting $a\cos x + b\sin x$ to $k\cos(x \pm \alpha)$ or $k\sin(x \pm \alpha)$, $k > 0$, α in any quadrant

Applying algebraic and trigonometric skills to functions: Identifying and sketching related functions

◆ Identifying and sketching a function after a transformation of the form $kf(x)$, $f(kx)$, $f(x) + k$, $f(x + k)$ or a combination of these

➤ Sketching the graph of $y = f^{-1}(x)$ given the graph of $y = f(x)$

➤ Sketching the inverse of a logarithmic or an exponential function.

➤ Completing the square in a quadratic expression where the coefficient of x^2 is non-unitary.

Applying algebraic and trigonometric skills to functions: Determining composite and inverse functions

◆ Determining a composite function given $f(x)$ and $g(x)$, where $f(x)$, $g(x)$ can be trigonometric, logarithmic, exponential or algebraic functions – including basic knowledge of domain and range

◆ $y = f^{-1}(x)$ of functions

➤ Knowing and using the terms domain and range

Applying geometric skills to vectors: Determining vector connections

◆ Determining the resultant of vector pathways in three dimensions

◆ Working with collinearity

◆ Determining the coordinates of an internal division point of a line

Applying geometric skills to vectors: Working with vectors

◆ Evaluating a scalar product given suitable information and determining the angle between two vectors

➤ Applying properties of the scalar product

➤ Using unit vectors **i**, **j**, **k** as a basis

UNIT 2: RELATIONSHIPS AND CALCULUS

Applying algebraic skills to solve equations: Solving algebraic equations

◆ Factorising a cubic polynomial expression with unitary x^3 coefficient

◆ Solving a cubic equation with unitary x^3 coefficient

◆ Given the nature of the roots of an equation, using the discriminant to find an unknown

contd

- ➢ Factorising a cubic or quartic polynomial expression with non-unitary coefficient of the highest power
- ➢ Solving a cubic or quartic equation with non-unitary coefficient of the highest power
- ➢ Solving quadratic inequalities, $ax^2 + bx + c \geqslant 0$ (or $\leqslant 0$)
- ➢ Finding the coordinates of the point(s) of intersection of a straight line and a curve or of two curves
- ➢ Simplifying a numerical expression using more than one of the laws of logarithms and exponents

Applying trigonometric skills to solve equations: Solving trigonometric equations

- ♦ Solving trigonometric equations in degrees, involving trigonometric formulae, in a given interval
- ➢ Solving trigonometric equations in degrees or radians, including those involving the wave function or trigonometric formulae or identities, in a given interval

Applying calculus skills of differentiation: Differentiating functions

- ♦ Differentiating an algebraic function which is, or can be simplified to, an expression in powers of x
- ♦ Differentiating $k\sin x$, $k\cos x$
- ➢ Differentiating a composite function using the chain rule

Applying calculus skills of differentiation: Using differentiation to investigate the nature and properties of functions

- ♦ Determining the equation of a tangent to a curve at a given point by differentiation
- ➢ Determining where a function is strictly increasing/decreasing
- ➢ Sketching the graph of an algebraic function by determining stationary points and their nature as well as intersections with the axes and behaviour of $f(x)$ for large positive and negative values of x

Applying calculus skills of integration: Integrating functions

- ♦ Integrating an algebraic function which is, or can be, simplified to an expression of powers of x
- ♦ Integrating functions of the form $f(x) = (x + q)^n$, $n \neq -1$
- ♦ Integrating functions of the form $f(x) = p\cos x$ and $f(x) = p\sin x$
- ➢ Integrating functions of the form $f(x) = (px + q)^n$, $n \neq -1$
- ➢ Integrating functions of the form $f(x) = p\cos(qx + r)$ and $f(x) = p\sin(qx + r)$
- ➢ Solving differential equations of the form $\frac{dy}{dx} = f(x)$

Applying calculus skills of integration: Using integration to calculate definite integrals

- ♦ Calculating definite integrals of polynomial functions with integer limits
- ➢ Calculating definite integrals of functions with limits which are integers, radians, surds or fractions

UNIT 3: APPLICATIONS

Applying algebraic skills to rectilinear shapes

- ♦ Using $m = \tan\theta$ to calculate a gradient or an angle
- ♦ Finding the equation of a line parallel to and a line perpendicular to a given line
- ➢ Determining whether or not two lines are perpendicular
- ➢ Using properties of medians, altitudes and perpendicular bisectors in problems involving the equation of a line and intersection of lines

Applying algebraic skills to circles

- ♦ Determining and using the equation of a circle
- ♦ Using properties of tangency in the solution of a problem
- ➢ Determining the intersection of circles or a line and a circle

Applying algebraic skills to sequences: Modelling situations using sequences

- ♦ Determining a recurrence relation from given information and using it to calculate a required term
- ♦ Finding and interpreting the limit of a sequence, where it exists

Applying calculus skills to optimisation and area: Applying differential calculus

- ♦ Determining the optimal solution for a given problem
- ➢ Determining the greatest/least values of a function on a closed interval
- ➢ Solving problems using rate of change

Applying calculus skills to optimisation and area: Applying integral calculus

- ♦ Finding the area between a curve and the x-axis
- ♦ Finding the area between a straight line and a curve or two curves
- ➢ Determining and using a function from a given rate of change and initial conditions

1 ALGEBRA

QUADRATICS: COMPLETING THE SQUARE
EXPRESSIONS AND FORMULAE

WHAT YOU SHOULD ALREADY KNOW: UNITARY COEFFICIENT OF x^2

A quadratic expression $x^2 + bx + c$ can be written in the form $(x + p)^2 + q$ using the method of **completing the square**. You have already used this to find **minimum** values of quadratic functions and to sketch their graphs.

ONLINE

Head to www.brightredbooks.net/N5Maths and take the Completing Squares test to check your prior knowledge and revise what you should know.

ONLINE

For a quick tutorial, watch 'Completing the Square' at www.brightredbooks.net/N5Maths

Example: 1

By expressing in the form $(x + p)^2 + q$, find the minimum value of $x^2 - 12x + 25$.

Solution:

$x^2 - 12x + 25$

$= (x - 6)^2 - 36 + 25$

$= (x - 6)^2 - 11$

The minimum value is -11, which occurs when $x = 6$.

This corresponds to the minimum turning point $(6, -11)$, for the graph of $y = x^2 - 12x + 25$:

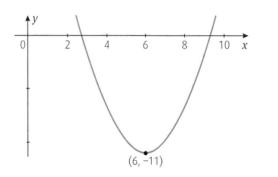

COMPLETING THE SQUARE: NON-UNITARY COEFFICIENT OF x^2

Completing the square in a quadratic expression where the coefficient of x^2 is non-unitary requires an additional initial step.

For expressions of the form $ax^2 + bx + c$, where $a \neq 1$, take out a common factor of a from the first two terms and then continue as before.

Example: 2

Express $2x^2 + 20x + 13$ in the form $a(x + p)^2 + q$.

Solution:

$2x^2 + 20x + 13 = 2[x^2 + 10x] + 13$ ——— take out a common factor of 2

$= 2[(x + 5)^2 - 25] + 13$ ——— complete the square for $[x^2 + 10x]$

$= 2(x + 5)^2 - 50 + 13 = 2(x + 5)^2 - 37$ ——— multiply out and complete

Example: 3

By expressing in the form $a(x + p)^2 + q$, find the maximum value of $14 - 12x - 3x^2$.

Solution:

$14 - 12x - 3x^2 = -3x^2 - 12x + 14$

$= -3[x^2 + 4x] + 14$

$= -3[(x + 2)^2 - 4] + 14$

$= -3(x + 2)^2 + 12 + 14$

$= -3(x + 2)^2 + 26$

$= 26 - 3(x + 2)^2$

Any number squared is never negative (≥ 0), so the **maximum** value this expression can have is 26, which occurs when $(x + 2) = 0$, or $x = -2$.

DON'T FORGET

You may find it easier to change the order of the terms in the expression.

contd

Alternative method: equating coefficients

If you are asked to express a quadratic expression in the form $a(x + p)^2 + q$, you can first **multiply out**:

$a(x + p)^2 + q = ax^2 + 2apx + ap^2 + q$

and then **equate coefficients** in x^2, x and the constant terms.

Example: 4

Express $4x^2 - 12x + 15$ in the form $a(x + p)^2 + q$.

Solution:

$a(x + p)^2 + q = ax^2 + 2apx + ap^2 + q$

Now we have:

$4x^2 - 12x + 15 = ax^2 + 2apx + ap^2 + q$

Equating terms in x^2: $\quad a = 4$

Equating terms in x: $\quad 2ap = -12 \Rightarrow p = \frac{-12}{2 \times 4} = \frac{-3}{2}$

and, equating constants: $ap^2 + q = 15 \Rightarrow q = 15 - 4 \times \left(\frac{-3}{2}\right)^2 = 6$

so $\quad\quad\quad\quad\quad\quad 4x^2 - 12x + 15 = 4\left(x - \frac{3}{2}\right)^2 + 6.$

ONLINE

For more on completing the square, follow the link at www.brightredbooks.net

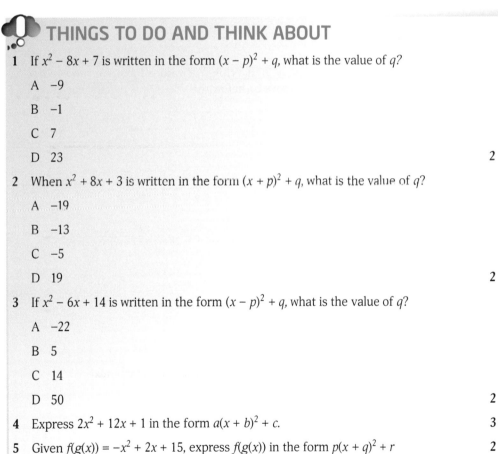

THINGS TO DO AND THINK ABOUT

1 If $x^2 - 8x + 7$ is written in the form $(x - p)^2 + q$, what is the value of q?

   A −9

   B −1

   C 7

   D 23   2

2 When $x^2 + 8x + 3$ is written in the form $(x + p)^2 + q$, what is the value of q?

   A −19

   B −13

   C −5

   D 19   2

3 If $x^2 - 6x + 14$ is written in the form $(x - p)^2 + q$, what is the value of q?

   A −22

   B 5

   C 14

   D 50   2

4 Express $2x^2 + 12x + 1$ in the form $a(x + b)^2 + c$.   3

5 Given $f(g(x)) = -x^2 + 2x + 15$, express $f(g(x))$ in the form $p(x + q)^2 + r$ 2

ONLINE TEST

Test yourself on quadratics at www.brightredbooks.net

QUADRATICS: SOLVING QUADRATIC INEQUALITIES ◄ EXPRESSIONS AND FORMULAE

SOLVING QUADRATIC INEQUALITIES

Solving a quadratic function, $f(x) = 0$, gives the x-coordinates of the points where the graph of the function cuts the x-axis. These x-values are called the roots of the quadratic function.

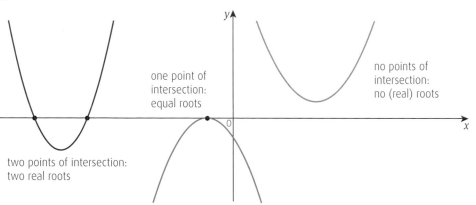

two points of intersection: two real roots

one point of intersection: equal roots

no points of intersection: no (real) roots

DON'T FORGET

To show that you are clearly solving an inequality by using >, ≥, ≤ or <.

When solving a **quadratic inequality**, we find a **range of values for x**. We can think of this as finding the part of the graph **above** or **below** the x-axis. It is usually easier to solve the corresponding equation first.

Example: 1

Solve $24 + 5x - x^2 > 0$.

Solution:

Sketching a graph can often help in solving quadratic inequalities.

$24 + 5x - x^2 = 0 \Rightarrow (8 - x)(3 + x) = 0$
when $x = 8$ or $x = -3$.

The parabola $y = 24 + 5x - x^2$ has negative x^2, so you know the shape.

We only need to know where $y > 0$. This is the range of values of x for which the graph is above the x-axis.

From the diagram it can be seen that $24 + 5x - x^2 > 0$ when $-3 < x < 8$.

Alternatively, instead of a sketch, take a value for x between $x = -3$ and $x = 8$ and substitute into an equation:

$y = 24 + 5x - x^2$

and evaluate to find y. If the value of y is positive $(24 + 5x - x^2 > 0)$, then the range required is between $x = -3$ and $x = 8$, as in this case, so $-3 < x < 8$.

$24 + 5x - x^2 > 0$

DON'T FORGET

If the part of the graph you want is continuous then the inequality is joined, e.g. $a < x < b$. If the part of the graph you want is separated then the inequalities are separate e.g. $x < a$ and $x > b$.

ONLINE

For more on inequations and inequalities, follow the links at www.brightredbooks.net

If $24 + 5x - x^2 < 0$ had been asked for in the example, then the required range of x would be outside this range so $x < -3$ and $x > 8$.

THINGS TO DO AND THINK ABOUT

1 Given $f(x) = 2x^2 + 5x - 12$, find

    (a) the range of x for which $f(x) < 0$.   2

    (b) the range of x which gives $f(x) \geqslant 0$.   1

2 For what values of x is $9x^2 - 12x + 4$ positive?   2

3 For what values of x is $12 + x - x^2 < 0$?

    A $x > 4$ only

    B $x < -3$ only

    C $x < -3, x > 4$

    D $-4 < x < 3$ 2

4 What is the solution of $x^2 + 5x > 0$, where x is a real number?

    A $-5 < x < 0$

    B $x < -5, x > 0$

    C $0 < x < 5$

    D $x < 0, x > 5$ 2

5 Solve $12 - x - x^2 < 0$.

    A $-4 < x < 3$

    B $x < -4, x > 3$

    C $-3 < x < 4$

    D $x < -3, x > 4$ 2

6 Solve $1 - 4x - 5x^2 > 0$, where x is a real number.

    A $x < -1$ or $x > \frac{1}{5}$

    B $-1 < x < \frac{1}{5}$

    C $x < -\frac{1}{5}$ or $x > 1$

    D $-\frac{1}{5} < x < 1$ 2

7 ABCD is a rectangle with sides of lengths x centimetres and $(x - 2)$ centimetres, as shown.

    If the area of ABCD is less than 15cm², determine the range of possible values of x.   4

8 If $f(x) = (x - 4)(x + 7)$, for what values of x is the graph of $y = f(x)$ above the x-axis?

    A $-7 < x < 4$

    B $-4 < x < 7$

    C $x < -7, x > 4$

    D $x < -4, x > 7$ 2

ONLINE TEST

Head to www.brightredbooks.net to take the test on quadratics.

QUADRATICS: THE DISCRIMINANT
RELATIONSHIPS AND CALCULUS

DON'T FORGET

The quadratic formula:

$x = \dfrac{-b \pm \sqrt{b^2 - 4ac}}{2a}$

DON'T FORGET

The nature of the roots of a quadratic equation depends on whether the discriminant is positive, negative or zero.

ONLINE

Head to www.brightredbooks.net/ N5Maths and take the Discriminant test to check your prior knowledge and revise what you should know.

ONLINE

A short recap on how to use the discriminant can be found at www.brightredbooks.net/ N5Maths

WHAT YOU SHOULD ALREADY KNOW

To solve quadratic equations of the form $ax^2 + bx + c = 0$, you will be familiar with using the quadratic formula $x = \dfrac{-b \pm \sqrt{b^2 - 4ac}}{2a}$.

At this level, the formula does not appear on the formulae list and you are expected to remember it.

The element under the square root sign, $b^2 - 4ac$, is the **discriminant**. It gives the nature of the roots of the quadratic equation.

You also know how to find an unknown when the resulting equation is linear.

The following two examples should remind you.

Example: 1

What is the nature of the roots of the equation $9x^2 - 24x + 16 = 0$?

Solution:

Using $b^2 - 4ac$, where $a = 9$, $b = -24$ and $c = 16$, we get $(-24)^2 - 4 \times 9 \times 16 = 0$.
So the roots are real and equal.

Example: 2

The equation $px^2 + 6x + 3 = 0$ has no real roots. Find the range of values of p.

Solution:

For no real roots $b^2 - 4ac < 0 \Rightarrow 6^2 - 4 \times p \times 3 < 0$
$36 - 12p < 0 \Rightarrow 36 < 12p \Rightarrow 3 < p$ and hence $p > 3$.

USING THE DISCRIMINANT: NATURE OF ROOTS AND SOLVING UNKNOWN(S) IN A NON-LINEAR EQUATION

At this level, your knowledge and skills in using the discriminant are extended.

You are expected to understand the terms **rational** and **irrational**.

You are also expected to find an unknown when the resulting equation is non-linear.

The following examples should help you understand what is required.

Example: 3

A function f is given by $f(x) = \frac{1}{4}x^2 + 3x + 7$. Describe the nature of the roots of $f(x) = 0$.

ONLINE

For more on irrational numbers, follow the links at www.brightredbooks.net

Solution:

$b^2 - 4ac = 3^2 - 4 \times \frac{1}{4} \times 7 = 2 > 0$.
$x = \dfrac{-b \pm \sqrt{b^2 - 4ac}}{2a} = \dfrac{-3 \pm \sqrt{2}}{\frac{1}{2}} = -6 \pm 2\sqrt{2}$.
The roots are real and distinct, but also irrational. $\sqrt{2}$ is an irrational number.

Example: 4

Find the range of the values of k such that $x^2 + kx - k + 3 = 0$ has real roots.

Solution:

We want $b^2 - 4ac \geq 0$ so $k^2 - 4(3 - k) \geq 0 \Rightarrow k^2 + 4k - 12 \geq 0$

contd

ONLINE

For more on the discriminant, follow the link at www.brightredbooks.net

Sketching a graph can help in solving a quadratic inequality like this.

$k^2 + 4k - 12 = 0$ when $k = -6$ or $k = 2$

The parabola $y = k^2 + 4k - 12$ has positive k^2 so you know the shape.

We want $k^2 + 4k - 12 \geq 0$, so we require $k = -6$ or $k = 2$ and the sections of the graph above the k-axis.

For $x^2 + kx - k + 3 = 0$ to have real roots, $k \leq -6$ or $k \geq 2$.

Alternatively, choose a value of k between $k = -6$ or $k = 2$ and evaluate $k^2 + 4k - 12$ for this value. For example, choose $k = 1$, and evaluate:

$k^2 + 4k - 12 = (1)^2 + 4(1) - 12 = -7 < 0$.

This does not give $k^2 + 4k - 12 \geq 0$ so the range required is not between 6 and 2 but outside it, so $k \leq -6$ or $k \geq 2$.

Tangents to a curve

Sometimes, it might not be so obvious that the discriminant is relevant in a question.

Example: 5

Show that the line $y = 5 - 4x$ does not intersect the parabola $y = x^2 - 6x + 7$.

Solution:

At a point of intersection, the line and the parabola would have the same coordinates, so

$x^2 - 6x + 7 = 5 - 4x$ equating the y-coordinates

leading to $x^2 - 2x + 2 = 0$ a quadratic equation

For this quadratic equation the discriminant $b^2 - 4ac = (-2)^2 - 4 \times 1 \times 2 = -4 < 0$

As $b^2 - 4ac < 0$, there are no real roots, and the line and the parabola do not intersect.

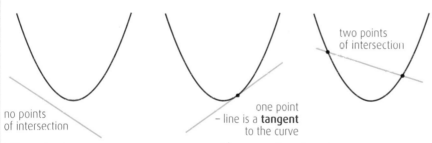

no points of intersection

one point – line is a **tangent** to the curve

two points of intersection

When the discriminant is not negative ($b^2 - 4ac \geqslant 0$), solve the quadratic (by factorising, using the quadratic formula or completing the square) to find the points of intersection.

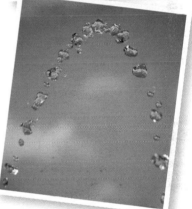

THINGS TO DO AND THINK ABOUT

1 Calculate the discriminant of the quadratic equation $2x^2 + 5x + 6 = 0$.

   A −38   B −23   C 58   D 73   **2**

2 The roots of the equation $kx^2 - 5x + 3 = 0$ are equal. What is the value of k?

   A $-\frac{25}{12}$ B $-\frac{12}{25}$ C $\frac{12}{25}$ D $\frac{25}{12}$ **2**

3 A function f is given by $f(x) = 3x^2 - x - 6$.

   Which of the following describes the nature of the roots of $f(x) = 0$?

   A No real roots   B Equal roots   C Real distinct roots   D Rational distinct roots   **2**

4 The discriminant of a quadratic equation is 17.

   Here are two statements about this quadratic equation:

   (1) the roots are real;   (2) the roots are rational.

   Which of the following is true?

   A Neither statement is correct.   B Only statement (1) is correct.
   C Only statement (2) is correct.   D Both statements are correct.   **2**

DON'T FORGET

$b^2 - 4ac < 0$ no real roots, line does not meet the parabola
$b^2 - 4ac = 0$ equal real roots, line is a tangent (just touches) the parabola
$b^2 - 4ac > 0$ two real distinct roots, line cuts the parabola twice

ONLINE TEST

Test yourself on quadratics at www.brightredbooks.net

POLYNOMIALS 1 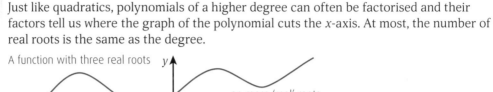 RELATIONSHIPS AND CALCULUS

Polynomial functions are of the form $ax^n + bx^{n-1} + cx^{n-2} + \ldots + kx^0$ where $a \neq 0$.

Quadratic functions are polynomials of **degree 2**.

The **cubic** function $f(x) = x^3 - 3x^2 - 10x + 24$ is a polynomial of **degree 3** (3 being the highest index of x).

The **quartic** function $g(x) = x^4 - x^2$ is a polynomial of **degree 4**.

ONLINE

Head to
www.brightredbooks.net
and follow the link for more
on the remainder theorem
and factor theorem.

DON'T FORGET

If $f(a) = 0$ then $x = a$ is a root.

DON'T FORGET

If $x = a$ is a root,
then $(x - a)$ is a factor.

DON'T FORGET

If $f(x)$ is divided by $(x - a)$
and there is no remainder
then $(x - a)$ is a factor
and $x = a$ is a root.

FACTORISING

Just like quadratics, polynomials of a higher degree can often be factorised and their factors tell us where the graph of the polynomial cuts the x-axis. At most, the number of real roots is the same as the degree.

A function with three real roots

no more 'real' roots

root   two equal roots

A function with three real roots.

The quadratic function $f(x) = x^2 - 5x + 6$ factorises to give $f(x) = (x - 2)(x - 3)$.

2 and 3 are factors of 6.

$f(2) = (2)^2 - 5(2) + 6 = 0$ and $f(3) = (3)^2 - 5(3) + 6 = 0$.

$x = 2$ and $x = 3$ are roots of $f(x) = 0$; $(x - 2)$ and $(x - 3)$ are factors of $f(x)$.

To factorise $f(x) = x^3 - 4x^2 + x + 6$, we first need to find a root. To do this we start with factors of 6, $\pm 1, \pm 2, \pm 3, \pm 6$ and substitute into $f(x)$ until we get zero with one of them.

$f(1) = (1)^3 - 4(1)^2 + (1) + 6 = 4 \neq 0$ so $x = 1$ is not a root.

$f(2) = (2)^3 - 4(2)^2 + (2) + 6 = 0$ so $x = 2$ is a root and so $(x - 2)$ is a factor.

The remainder when $f(x)$ is divided by $(x - 2)$ is $f(2)$. Since the remainder is **zero** we have found a factor. For a function $f(x)$, $f(a)$ gives the remainder when $f(x)$ is divided by $(x - a)$. Below we look at methods of division.

It would be possible to find other factors of 6 which are also roots of $f(x) = 0$ and hence fully factorise $f(x)$.

For example

$f(3) = (3)^3 - 4(3)^2 + (3) + 6 = 0$ so $x = 3$ is a root and $(x - 3)$ is a factor.

$f(-1) = (-1)^3 - 4(-1)^2 + (-1) + 6 = 0$ so $x = -1$ is a root and $(x + 1)$ is a factor.

Thus $f(x) = x^3 - 4x^2 + x + 6 = (x + 1)(x - 2)(x - 3)$.

However, generally once one factor is found then we divide to find others.

There are three common ways to divide:

- long division
- synthetic division
- by inspection.

Examples in this chapter will demonstrate the three methods, but you should stick to the one you are happiest using.

Example: 1

$(x - 2)$ is a factor of $f(x) = x^3 + px - 2$. Find the value of p and factorise the cubic expression fully.

contd

Solution:

Since $(x - 2)$ is a factor, $x = 2$ is a root and $f(2) = 0$.

$f(2) = (2)^3 + p(2) - 2 = 0 \Rightarrow 6 + 2p = 0 \Rightarrow p = -3$

Using **long division** we now divide $x^3 + 0x^2 - 3x - 2$ by $(x - 2)$.

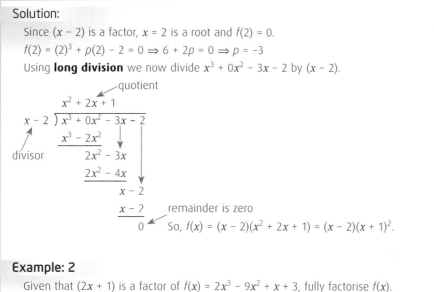

So, $f(x) = (x - 2)(x^2 + 2x + 1) = (x - 2)(x + 1)^2$.

DON'T FORGET

If there are any missing powers of x, set their coefficient = 0.

ONLINE

For more on algebraic long division, follow the link at www.brightredbooks.net

Example: 2

Given that $(2x + 1)$ is a factor of $f(x) = 2x^3 - 9x^2 + x + 3$, fully factorise $f(x)$.

ONLINE

For more on synthetic division, follow the link at www.brightredbooks.net

Solution:

By **synthetic division**

We need to rewrite $(2x + 1)$ as $2\left(x + \frac{1}{2}\right)$ and divide $f(x)$ by $\left(x + \frac{1}{2}\right)$.

remainder = 0

So, $f(x) = \left(x + \frac{1}{2}\right)(2x^2 - 10x + 6)$ ... but this now needs some rearrangement:

$f(x) = \left(x + \frac{1}{2}\right) \times 2(x^2 - 5x + 3)$ ... removing the factor 2 from the second bracket

$= (2x + 1)(x^2 - 5x + 3)$ ... and putting it back in the first,

so the other factor is $(x^2 - 5x + 3)$. This does not factorise.

$(b^2 - 4ac = 13$ and $\sqrt{13}$ is irrational.)

Example: 3

A function f is defined on the set of real numbers by $f(x) = x^3 - 7x^2 + 2x + 40$.

Show that when $f(x)$ is divided by $(x - 4)$ the remainder is zero. Hence fully factorise $f(x)$.

Solution:

If the remainder is zero when $f(x)$ is divided by $(x - 4)$ then $f(4) = 0$.

Substituting $x = 4$ into $f(x) = x^3 - 7x^2 + 2x + 40$ gives

$f(4) = (4)^3 - 7(4)^2 + 2(4) + 40 = 64 - 112 + 8 + 40 = 0$, hence the remainder is zero.

$(x - 4)$ is a factor of $f(x)$.

By **inspection**

$f(x) = x^3 - 7x^2 + 2x + 40 = (x - 4)(x^2 - 3x - 10) = (x - 4)(x - 5)(x + 2)$

required for x^3

$\underset{2}{\overset{1}{(x - 4)}} \quad (x^2 - 3x - 10)$

$\overset{3}{-4x^2}$ $\quad$ $\overset{4}{}$ check: $-4 \times -10 = 40$

need $-7x^2$

$(-3x^2$ required$)$

$12x$

need $2x$ $(-10x$ required$)$

ONLINE

For more on solving polynomials, follow the link at www.brightredbooks.net

ONLINE TEST

Head to www.brightredbooks.net to test yourself on polynomials.

THINGS TO DO AND THINK ABOUT

1 Given that $(x + 1)$ is a factor of $x^3 + 3x^2 + 5x + 3$, factorise this cubic fully. $\qquad$ 4

POLYNOMIALS 2 — RELATIONSHIPS AND CALCULUS

DON'T FORGET

Once a polynomial has been factorised it is just one further step to solve.

SOLVING ALGEBRAIC EQUATIONS

If $f(x) = (x - 2)(x + 1)^2$, solving $f(x) = 0$ gives $(x - 2)(x + 1)^2 = 0$ and solutions $x = 2$ and $x = -1$.

Example: 1

A function f is defined on the set of real numbers by $f(x) = x^3 - x^2 - 9x + 9$.

(a) Show that $x = 1$ is a root of $f(x) = 0$.

(b) Hence fully factorise $f(x)$.

(c) Find the solutions of $x^3 - x^2 - 9x + 9 = 0$.

Solution:

(a) If $x = 1$ is a root of $f(x) = x^3 - x^2 - 9x + 9$ then $f(1) = 0$.

$f(1) = (1)^3 - (1)^2 - 9(1) + 9 = 0$, hence $x = 1$ is a root and $(x - 1)$ is a factor.

(b) By inspection:

$f(x) = x^3 - x^2 - 9x + 9 = (x - 1)(x^2 - 9) = (x - 1)(x - 3)(x + 3)$

(c) $x^3 - x^2 - 9x + 9 = 0 \Rightarrow (x - 1)(x - 3)(x + 3) = 0$ to give solutions $x = 1$, $x = 3$ and $x = -3$.

GRAPHS OF POLYNOMIAL FUNCTIONS

Solving an algebraic function gives us **intersection points**. In the example above, when $f(x) = x^3 - x^2 - 9x + 9$, solving $f(x) = 0$ gives the roots, which are the x-coordinates of the intersection with the x-axis. The graph of $f(x) = x^3 - x^2 - 9x + 9$ crosses the x-axis at $(-3, 0)$, $(1, 0)$ and $(3, 0)$.

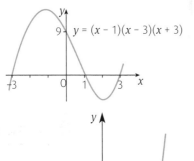

$x^3 - x^2 - 9x + 9$ has a positive coefficient of x^3 so it starts with a positive gradient and will pass through $(0, 9)$. ($f(0) = 9$.)

The graph of $f(x) = (x - 2)(x + 1)^2$ will also start with a positive gradient, but $x = -1$ is a repeated root, so the graph will just touch the x-axis at this point. It will cross the y-axis at $(0, -2)$ and through calculus (see page 44) you will find the minimum turning point is $(1, -4)$.

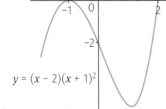

Investigating what happens to y as x becomes very large positive or negative ($x \rightarrow \pm \infty$) helps to verify the shape of a polynomial graph.

In both cases above:

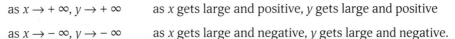

as $x \rightarrow + \infty$, $y \rightarrow + \infty$ as x gets large and positive, y gets large and positive

as $x \rightarrow - \infty$, $y \rightarrow - \infty$ as x gets large and negative, y gets large and negative.

Finding a polynomial function from knowledge of its graph

Example: 2

Find an expression for $f(x)$, the cubic function whose roots are $x = -5$, $x = -2$ and $x = 1$ and which cuts the y-axis at $(0, -20)$.

Solution:

The roots $x = -5$, $x = -2$ and $x = 1$ mean there are factors $(x + 5)$, $(x + 2)$ and $(x - 1)$, so

$f(x) = k(x + 5)(x + 2)(x - 1)$.

Using $(0, -20)$: $-20 = k(0 + 5)(0 + 2)(0 - 1)$, so $k = 2$ and

$f(x) = 2(x + 5)(x + 2)(x - 1) = 2x^3 + 12x^2 + 6x - 20$.

GRAPHS OF POLYNOMIAL FUNCTIONS

Finding the intersection of a curve and a straight line or two curves

A straight line may meet the graph of a cubic function three times, twice, once or not at all.

Two graphs of polynomial functions could have several points of intersection (or none at all). Where they exist, they can be found, just as we found points of intersection of a line and a parabola earlier.

one point of intersection

two points, crosses once and just touches

three points of intersection

Example: 3

Find all the points of intersection of the curves $y = -x^3 + 12x + 4$ and $y = x^2 + 6x + 4$.

Solution:

Equate expressions for y and rearrange to form a cubic:

$x^2 + 6x + 4 = -x^3 + 12x + 4$

$x^3 + x^2 - 6x = 0$

$x(x^2 + x - 6) = 0$

$x(x - 2)(x + 3) = 0$, so $x = 0$, $x = 2$, $x = -3$

Substituting these x-values into the equation of either curve gives the corresponding y-values and hence the coordinates of points of intersection: $(-3, -5)$, $(0, 4)$ and $(2, 20)$.

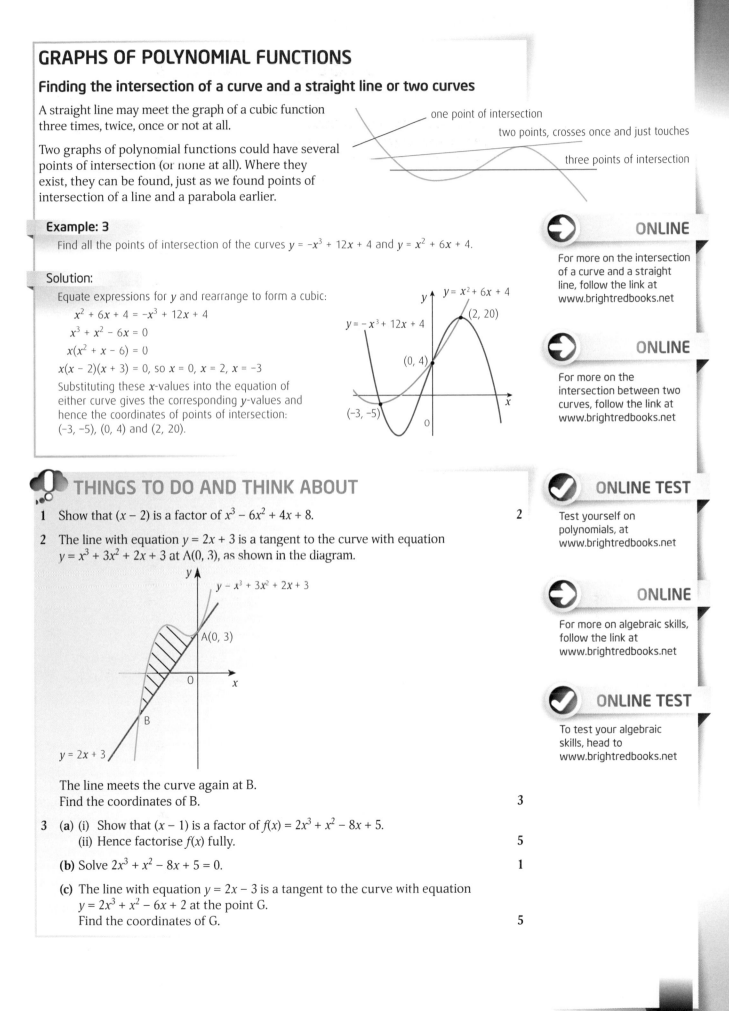

ONLINE

For more on the intersection of a curve and a straight line, follow the link at www.brightredbooks.net

ONLINE

For more on the intersection between two curves, follow the link at www.brightredbooks.net

THINGS TO DO AND THINK ABOUT

1 Show that $(x - 2)$ is a factor of $x^3 - 6x^2 + 4x + 8$. **2**

2 The line with equation $y = 2x + 3$ is a tangent to the curve with equation $y = x^3 + 3x^2 + 2x + 3$ at A$(0, 3)$, as shown in the diagram.

The line meets the curve again at B.
Find the coordinates of B. **3**

3 (a) (i) Show that $(x - 1)$ is a factor of $f(x) = 2x^3 + x^2 - 8x + 5$.
    (ii) Hence factorise $f(x)$ fully. **5**

  (b) Solve $2x^3 + x^2 - 8x + 5 = 0$. **1**

  (c) The line with equation $y = 2x - 3$ is a tangent to the curve with equation $y = 2x^3 + x^2 - 6x + 2$ at the point G.
    Find the coordinates of G. **5**

ONLINE TEST

Test yourself on polynomials, at www.brightredbooks.net

ONLINE

For more on algebraic skills, follow the link at www.brightredbooks.net

ONLINE TEST

To test your algebraic skills, head to www.brightredbooks.net

FUNCTIONS: DOMAIN AND RANGE
EXPRESSIONS AND FUNCTIONS

FUNCTIONS

A **function** is a rule which **maps** every member of a set, say A, to **exactly one member** of a set B.

The set of input values is called the **domain** of the function.

If a belongs to A and $f(a) = b$, then b is called the **image** (the output) of a.

The set of all images (output values) is called the **range** of the function.

The **codomain** of a function is the set into which all of the ouput values fall.

In set notation $f: A \rightarrow B$, A is the **domain**, B is the **codomain**.

Remember your symbols

You are expected to be able to understand the use of the content below.

Symbols, terms and sets:

the symbols: $\in$, $\notin$, { } or $\emptyset$

the terms: set, subset, empty set, member, element

the conventions for representing sets, namely:

N, the set of natural numbers, {1, 2, 3, ...}

N₀, the set of whole numbers, {0, 1, 2, 3, ...}, formerly **W**

Z, the set of integers

Q, the set of rational numbers

R, the set of real numbers

$\in$ means 'belongs to' or 'is a member of'

$\notin$ means 'does not belong to'

{ } or $\emptyset$ is used to denote an empty set

$\Leftrightarrow$ means 'if and only if'

Rational numbers can be expressed in the form $\frac{m}{n}$, where m and n are integers, $n \neq 0$.

Irrational numbers cannot be expressed in the form $\frac{m}{n}$. An example of an irrational number is $\sqrt{2}$.

ONLINE

Head to
www.brightredbooks.net
for more on numbers.

Example: 1

What is the range for $f(x) = x^2$, where $f: \mathbf{R} \rightarrow \mathbf{R}$, and for $h(x) = 2x + 1$, where $h: \mathbf{R} \rightarrow \mathbf{R}$?

Solution:

$f(x) = x^2$, **R** is the domain, it is also the codomain, but the range is a subset of the codomain.

Consider the input, 2 and –2: $f(2) = 4$ and $f(-2) = 4$.

The range, the set of all images, is all real numbers less negative numbers.

However, for the function h, where $h(x) = 2x + 1$, and $h: \mathbf{R} \rightarrow \mathbf{R}$, the codomain and the range are the same: they are the set of real numbers, **R**.

contd

Solution:

Rewriting $f(x)$ as $\frac{\bullet}{1 + \bullet}$ gives a clear picture as to where to put the output of your first function.

$$f(f(x)) = \frac{\frac{1}{1+x}}{1 + \frac{1}{1+x}}$$

Take the denominator and make it a single fraction:

$$1 + \frac{1}{1+x} = \frac{1+x}{1+x} + \frac{1}{1+x} = \frac{2+x}{1+x}$$

$$\Rightarrow f(f(x)) = \frac{\frac{1}{1+x}}{\frac{2+x}{1+x}} = \frac{1}{1+x} \times \frac{1+x}{2+x} = \frac{1}{2+x}$$

THINGS TO DO AND THINK ABOUT

1 The functions f and g are defined by $f(x) = x^2 + 1$ and $g(x) = 3x - 4$, on the set of real numbers.

Find $g(f(x))$. **2**

2 Functions f and g are defined on the set of real numbers by

$f(x) = x^2 + 3$ $\qquad$ $g(x) = x + 4$

Find expressions for

(a) $f(g(x))$, **2**

(b) $g(f(x))$. **1**

3 Functions f and g are defined on suitable domains by

$f(x) = x(x - 1) + q$ and $g(x) = x + 3$.

Find an expression for $f(g(x))$. **2**

4 Functions f and g are defined on suitable domains by $f(x) = \cos x$ and $g(x) = x + \frac{\pi}{6}$.

What is the value of $f\left(g\left(\frac{\pi}{6}\right)\right)$? **3**

5 Functions f, g and h are defined on the set of real numbers by

$f(x) = x^3 - 1$ $\qquad$ $g(x) = 3x + 1$ $\qquad$ $h(x) = 4x - 5$.

(a) Find $g(f(x))$. **2**

(b) Show that $g(f(x)) + xh(x) = 3x^3 + 4x^2 - 5x - 2$. **1**

6 The functions f and g are defined by

$f(x) = 4x + 3$, $x \in \mathbf{R}$, and $g(x) = \frac{1}{x^2}$, $x \in \mathbf{R}$, $x \neq 0$.

Write down

(a) the composite function $f(g(x))$, **1**

(b) the inverse function $f^{-1}(x)$. **2**

VIDEO LINK

Check out the clip about composite functions at www.brightredbooks.net

ONLINE

For more on this, follow the link at www.brightredbooks.net

ONLINE TEST

Test yourself on your knowledge of this topic online at www.brightredbooks.net

FUNCTIONS: GRAPHS OF FUNCTIONS 1
EXPRESSIONS AND FUNCTIONS

DON'T FORGET +

Graphic calculators and
graphing packages are
very useful for investigating
and checking.

DON'T FORGET +

You need to be able
to sketch and identify
functions without a
calculator.

TRANSFORMATIONS

Given the graph of a function $f(x)$, you will be expected to be able to identify and sketch related functions.

The table describes basic transformations and the examples below show combinations of these.

$y = -f(x)$	Graph is reflected in the x-axis	$(x, y) \rightarrow (x, -y)$ $(0, 0) \rightarrow (0, 0)$	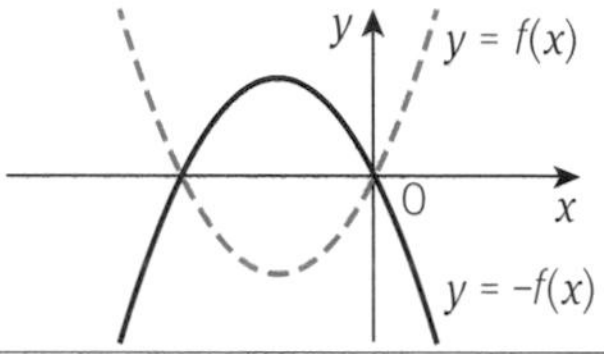
$y = f(-x)$	Graph is reflected in the y-axis	$(x, y) \rightarrow (-x, y)$ $(0, 0) \rightarrow (0, 0)$	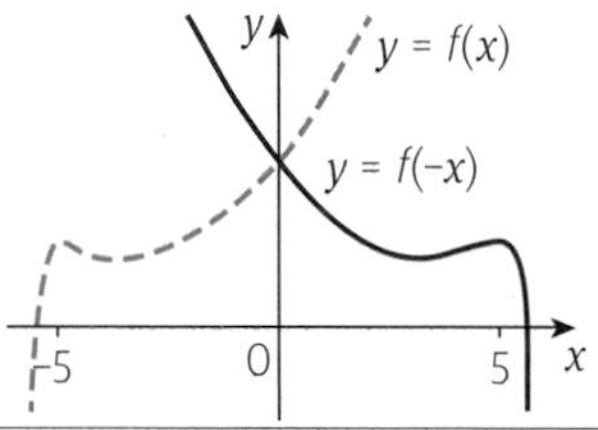
$y = af(x)$	Graph is stretched (or compressed) parallel to the y-axis (vertically) by a factor of a	$(x, y) \rightarrow (x, ay)$ $(0, 0) \rightarrow (0, 0)$	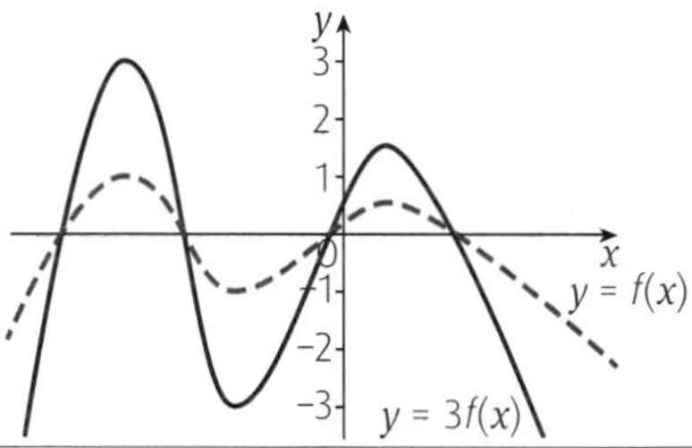
$y = f(bx)$	Graph is compressed (or stretched) parallel to the x-axis (horizontally) by a factor of b	$(x, y) \rightarrow (\frac{x}{b}, y)$ $(0, 0) \rightarrow (0, 0)$	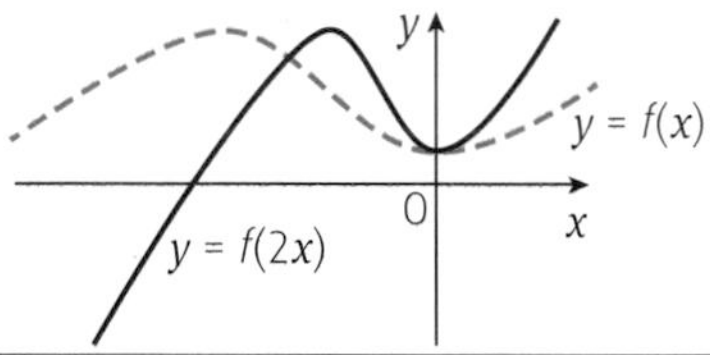
$y = f(x) + c$	Graph slides parallel to the y-axis: up $(c > 0)$ or down $(c < 0)$	$(x, y) \rightarrow (x, y + c)$ $(0, 0) \rightarrow (0, c)$	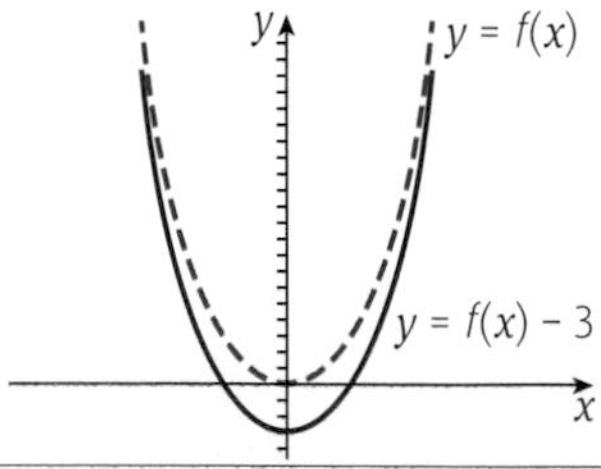
$y = f(x + d)$	Graph slides parallel to the x-axis: to the left $(d > 0)$ or to the right $(d < 0)$	$(x, y) \rightarrow (x + d, y)$ $(0, 0) \rightarrow (-d, 0)$	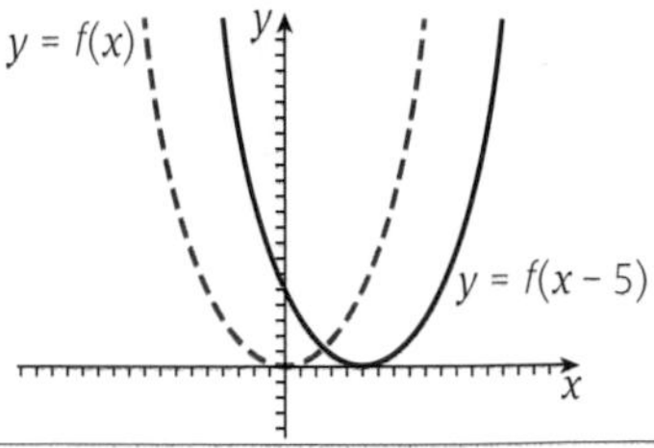

Example: 1

Sketch $y = x^2$ and on the same diagram sketch $y = -\frac{1}{2}(x + 3)^2 + 2$

contd

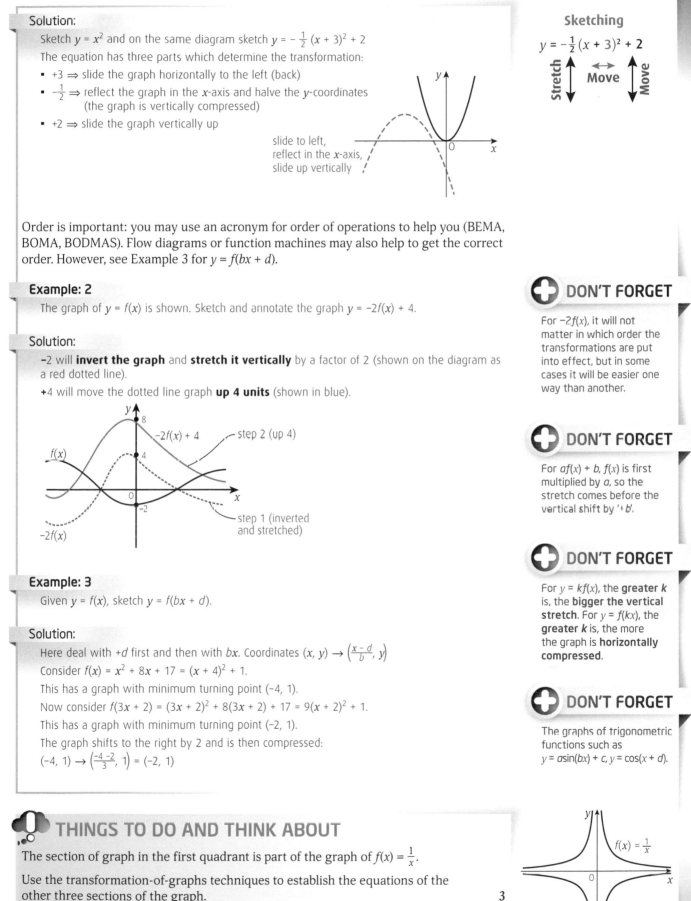

Solution:

Sketch $y = x^2$ and on the same diagram sketch $y = -\frac{1}{2}(x + 3)^2 + 2$

The equation has three parts which determine the transformation:

- $+3 \Rightarrow$ slide the graph horizontally to the left (back)
- $-\frac{1}{2} \Rightarrow$ reflect the graph in the x-axis and halve the y-coordinates (the graph is vertically compressed)
- $+2 \Rightarrow$ slide the graph vertically up

slide to left,
reflect in the x-axis,
slide up vertically

Order is important: you may use an acronym for order of operations to help you (BEMA, BOMA, BODMAS). Flow diagrams or function machines may also help to get the correct order. However, see Example 3 for $y = f(bx + d)$.

Example: 2

The graph of $y = f(x)$ is shown. Sketch and annotate the graph $y = -2f(x) + 4$.

Solution:

-2 will **invert the graph** and **stretch it vertically** by a factor of 2 (shown on the diagram as a red dotted line).

$+4$ will move the dotted line graph **up 4 units** (shown in blue).

Example: 3

Given $y = f(x)$, sketch $y = f(bx + d)$.

Solution:

Here deal with $+d$ first and then with bx. Coordinates $(x, y) \rightarrow \left(\frac{x - d}{b}, y\right)$

Consider $f(x) = x^2 + 8x + 17 = (x + 4)^2 + 1$.

This has a graph with minimum turning point $(-4, 1)$.

Now consider $f(3x + 2) = (3x + 2)^2 + 8(3x + 2) + 17 = 9(x + 2)^2 + 1$.

This has a graph with minimum turning point $(-2, 1)$.

The graph shifts to the right by 2 and is then compressed:

$(-4, 1) \rightarrow \left(\frac{-4 - 2}{3}, 1\right) = (-2, 1)$

Sketching

$$y = -\frac{1}{2}(x + 3)^2 + 2$$

Stretch ↕ ↔ Move ↕ Move

➕ **DON'T FORGET**

For $-2f(x)$, it will not matter in which order the transformations are put into effect, but in some cases it will be easier one way than another.

➕ **DON'T FORGET**

For $af(x) + b$, $f(x)$ is first multiplied by a, so the stretch comes before the vertical shift by '$+ b$'.

➕ **DON'T FORGET**

For $y = kf(x)$, the **greater k** is, the **bigger the vertical stretch**. For $y = f(kx)$, the greater k is, the more the graph is **horizontally compressed**.

➕ **DON'T FORGET**

The graphs of trigonometric functions such as $y = a\sin(bx) + c$, $y = \cos(x + d)$.

🗯 THINGS TO DO AND THINK ABOUT

The section of graph in the first quadrant is part of the graph of $f(x) = \frac{1}{x}$.

Use the transformation-of-graphs techniques to establish the equations of the other three sections of the graph.

3

$f(x) = \frac{1}{x}$

FUNCTIONS: GRAPHS OF FUNCTIONS 2
EXPRESSIONS AND FUNCTIONS

DON'T FORGET

Not all functions have inverses.

DON'T FORGET

Use the same scale on both axes.

SKETCHING THE INVERSE FUNCTION

To sketch the graph of $y = f^{-1}(x)$ given the graph of $y = f(x)$, reflect $y = f(x)$ in the line $y = x$.

The coordinates switch $(x, y) \rightarrow (y, x)$.

Provided the scale is the same on both axes, you will see that $y = x$ is a **line of symmetry**.

Example: 1

Given the graph of $y = f(x)$ (shown in blue), sketch $y = f^{-1}(x)$.

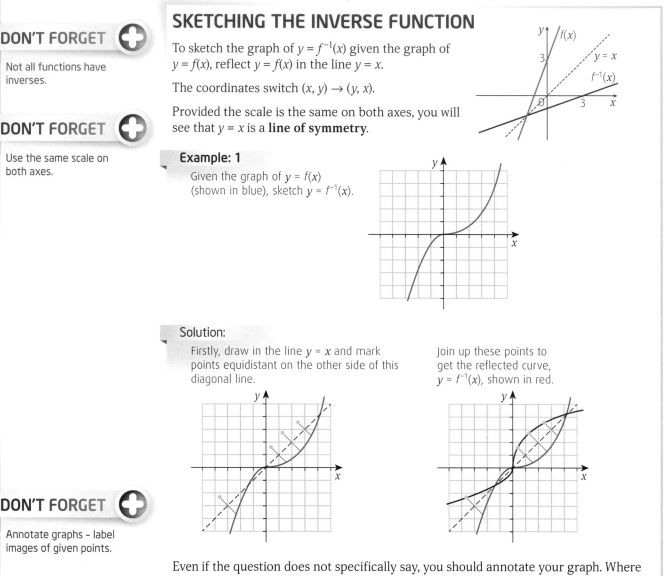

Solution:

Firstly, draw in the line $y = x$ and mark points equidistant on the other side of this diagonal line.

Join up these points to get the reflected curve, $y = f^{-1}(x)$, shown in red.

DON'T FORGET

Annotate graphs – label images of given points.

Even if the question does not specifically say, you should annotate your graph. Where points of the original graph have been labelled, or given, you should label the image of those points.

SKETCHING THE GRAPHS OF EXPONENTIAL AND LOGARITHMIC FUNCTIONS

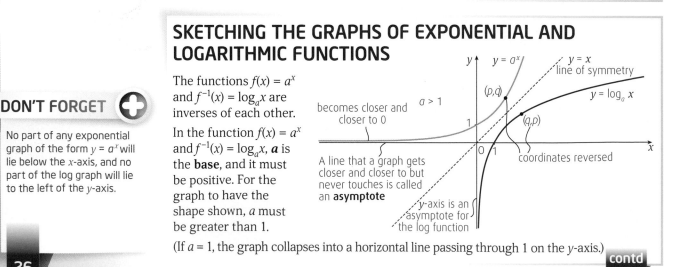

DON'T FORGET

No part of any exponential graph of the form $y = a^x$ will lie below the x-axis, and no part of the log graph will lie to the left of the y-axis.

The functions $f(x) = a^x$ and $f^{-1}(x) = \log_a x$ are inverses of each other.

In the function $f(x) = a^x$ and $f^{-1}(x) = \log_a x$, a is the **base**, and it must be positive. For the graph to have the shape shown, a must be greater than 1.

becomes closer and closer to 0

A line that a graph gets closer and closer to but never touches is called an **asymptote**

y-axis is an asymptote for the log function

(If $a = 1$, the graph collapses into a horizontal line passing through 1 on the y-axis.)

contd

Different values of a will affect the steepness of the slope. The lines that the graphs approach but never touch are **asymptotes**, and the points $(1, 0)$ on $y = \log_a x$ and $(0, 1)$ on $y = a^x$, are unaffected by changes in the value of a.

To sketch $y = a^{-x}$, we reflect $y = a^x$ in the y-axis.

Example: 2

Sketch $y = \left(\frac{1}{2}\right)^x$.

Solution:

$y = \left(\frac{1}{2}\right)^x = 2^{-x}$, so sketch $y = 2^x$ and reflect in the y-axis.

$y = 2^{-x}$

or $y = \left(\frac{1}{2}\right)^x = 2^{-x}$

$y = 2^x$

DON'T FORGET

Remember $y = a^{-x} = \left(\frac{1}{a}\right)^x$.

Taking a point on the exponential graph and its image on the logarithmic graph (or vice versa):

(p, q) lies on $y = a^x$ (q, p) lies on $y = \log_a x$

Substituting in the coordinates: $q = a^p$ $p = \log_a q$

The two statements are equivalent – they are saying the same thing in different ways.

$$y = \log_a x \Leftrightarrow x = a^y$$

Example: 3

If $3 = \log_a 125$, find a.

Solution:

$3 = \log_a 125 \Rightarrow 125 = a^3$, but $5^3 = 125$
so $a = 5$

Example: 4

Evaluate $\log_2 16$

Solution:

You might find it helpful to think '2 to what power gives 16?'

Since $2 \times 2 \times 2 \times 2 = 16$, the answer is 4.

Example: 5

What is p, if $p = \log_3 \frac{1}{243}$?

Solution:

This is not an obvious one, so it might be helpful to rewrite it as an exponential:

$3^p = \frac{1}{243} = \frac{1}{3 \times 3 \times 3 \times 3 \times 3} = 3^{-4} \Rightarrow p = -4$

The function $\mathbf{y = e^x}$ is a special exponential function, where $e = \lim\limits_{n \to \infty} \left(1 + \frac{1}{n}\right)^n = 2{\cdot}71828$ (to 6 s.f.).

The inverse of $y = e^x$ is $y = \log_e x$ or $y = \ln x$, the **natural logarithm**.

On your calculator **LOG** or **log** is $\mathbf{log_{10}}$ and **LN** or **ln** is $\mathbf{log_e}$. For example:

$y = e^5 \Leftrightarrow \log_e y = 5$

$\ln y = \frac{1}{2} \Rightarrow \log_e y = \frac{1}{2} \Leftrightarrow y = e^{\frac{1}{2}} = \sqrt{e}$

ONLINE

Head to www.brightredbooks.net for some great links on this topic.

ONLINE TEST

Test yourself on your knowledge of this topic online at www.brightredbooks.net

THINGS TO DO AND THINK ABOUT

1 The diagram shows the graph of $y = f(x)$ where f is a logarithmic function.

$y = f(x)$

$(6, 1)$

$(4, 0)$

What is $f(x)$?

2

FUNCTIONS: EXPONENTIALS AND LOGARITHMIC FUNCTIONS

EXPRESSIONS AND FUNCTIONS

EXPONENTIALS AND LOGARITHMIC FUNCTIONS

$$y = \log_a x \Leftrightarrow x = a^y$$

For example:

$$10^3 = 1000 \Leftrightarrow \log_{10}1000 = 3$$

$$2 = \log_5 25 \Leftrightarrow 25 = 5^2$$

$$27^{\frac{1}{3}} = 3 \Leftrightarrow \log_{27}3 = \frac{1}{3}$$

A **logarithm** is an **index**:

$$y = \log_a x \Leftrightarrow x = a^y, \therefore x = a^{\log_a x}$$

and so we can apply the rules of indices.

$$\log_a x + \log_a y = \log_a xy$$

$$\log_a x - \log_a y = \log_a \frac{x}{y}$$

$$n\log_a x = \log_a x^n$$

$$\log_a 1 = 0$$

$$\log_a a = 1$$

Example: 1

$$\log_6 4 + \log_6 9 = \log_6 36 = 2$$

Example: 2

$$4\log_e \sqrt{e} = 4\log_e e^{\frac{1}{2}} = 2\log_e e = 2$$

Example: 3

$$\log_a \sqrt{x} + \frac{3}{2}\log_a x = \log_a 25$$

$$\Rightarrow \log_a x^{\frac{1}{2}} + \log_a x^{\frac{3}{2}} = \log_a 25$$

$$\Rightarrow \log_a x^2 = \log_a 25$$

$$\Rightarrow x = 5$$

USING LOGARITHMS TO ESTABLISH RELATIONSHIPS OF THE FORM $y = kx^n$ AND $y = ab^x$

Example: 4

x and y are related by the law $y = kx^n$ and the graph shown. Find k and n.

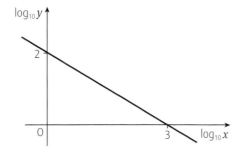

Solution:

Working from the equation:

$$y = kx^n$$

$$\log_{10} y = \log_{10} kx^n \quad \text{—— taking logs of both sides}$$

$$\log_{10} y = \log_{10} k + n\log_{10} x \quad \text{—— using } \log_a xy = \log_a x + \log_a y \text{ and } \log_a x^n = n\log_a x$$

$$\log_{10} y = n\log_{10} x + \log_{10} k$$

Working from the graph:

y-intercept = 2

gradient = $-\frac{2}{3}$

equation: $\log_{10} y = -\frac{2}{3}\log_{10} x + 2$

(noticing the labels on the axes)

Since these two formulas represent the same relationship, we can equate the coefficients of the linear and constant term:

$$n = -\frac{2}{3}$$

$$\log_{10} k = 2$$

$$k = 10^2 = 100$$

The formula becomes $y = 100x^{-\frac{2}{3}}$

contd

Example: 5

The graph of $\log_{10} y$ against x has been drawn for some experimental data. It is known that the data fits an equation $y = ab^x$ for some constants a and b. Find the values of a and b.

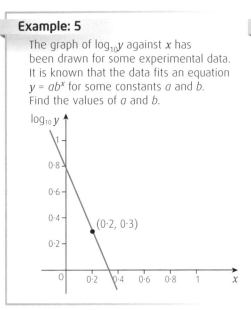

Solution:

Working from the equation:

$y = ab^x$

$\log_{10} y = \log_{10} ab^x$

$\log_{10} y = \log_{10} a + x \log_{10} b$

$\log_{10} y = (\log_{10} b) x + \log_{10} a$

Working from the graph:

y-intercept = 0·8

gradient $= \frac{0·3 - 0·8}{0·2 - 0} = -2·5$

equation: $\log_{10} y = -2·5x + 0·8$

Equating coefficients:

$\log_{10} b = -2·5 \Rightarrow b = 10^{-2·5}$

and: $\log_{10} a = 0·8 \Rightarrow a = 10^{0·8}$

and the formula is $y = 10^{0·8} (10^{-2·5})^x$

DON'T FORGET

$\log x$ and $\log y$ on the axes:
Linear law is
$\log y = n \log x + \log k$
Formula is $y = kx^n$

ONLINE

To learn more about logarithms, follow the links at www.brightredbooks.net

DON'T FORGET

x and $\log y$ on the axes:
Linear law is
$\log y = \log a + x \log b$
Formula is $y = ab^x$

EXPONENTIAL GROWTH AND DECAY

Real-life problems such as the spread of diseases, radioactive decay and financial investments can be modelled and solved by exponential and logarithmic functions.

Example: 6

The amount A_t, micrograms, of a radioactive substance remaining after t years decreases according to the formula $A_t = A_0 e^{-0·003t}$ where A_0 is the amount present initially.

(a) If 250 micrograms are left after 500 years, how many micrograms were present initially?

(b) Find the half-life of the substance.

Solution:

(a) We are told that $t = 500$ and $A_{500} = 250$

So $A_{500} = A_0 e^{-0·003 \times 500} = 250$

$\Rightarrow A_0 = \frac{250}{e^{-1·5}}$

Using a calculator gives $A_0 = 1120$ micrograms.

(b) $A_t = \frac{1}{2} A_0$.

Using $A_t = A_0 e^{-0·003t}$, we get $\frac{1}{2} A_0 = A_0 e^{-0·003t}$; so $\frac{1}{2} = e^{-0·003t}$.

Taking logs of both sides gives $\log_e \frac{1}{2} = \log_e e^{-0·003t} \Rightarrow \log_e \frac{1}{2} = -0·003t \log_e e$

$\Rightarrow \log_e \frac{1}{2} = -0·003t$ and $t = \frac{\log_e 0·5}{-0·003} = 231$

The half-life is 231 years.

DON'T FORGET

The half-life is the time it takes for the amount of substance to fall to half of its original amount.

Example: 7

The number of people in a city infected with a virus is increasing according to the formula $V_t = V_0 e^{0·343t}$ where t is the number of days since the outbreak. Initially, at 9am on 1 November, 10 people in the city are known to be infected.

(a) Calculate the number of people infected after one week.

(b) Hospital beds to treat infected people are limited to 1000. During which day will the hospital run out of beds?

Solution:

(a) After one week, $t = 7 \Rightarrow V_7 = 10e^{0·343 \times 7}$. Using a calculator $V_7 = 110$.

One hundred and ten people are infected.

(b) $V_t > 1000$ for the hospital to run out of beds.

$\Rightarrow 10e^{0·343t} > 1000 \Rightarrow e^{0·343t} > 100 \Rightarrow \log_e e^{0·343t} > \log_e 100$

$\Rightarrow 0·343t \log_e e > \log_e 100$,

therefore $t > \frac{\log_e 100}{0·343} \Rightarrow t > 13·4$

The hospital will run out of beds during 14 November.

THINGS TO DO AND THINK ABOUT

1   Solve the equation $\log_5(3 - 2x) + \log_5(2 + x) = 1$, where x is a real number.   4

FUNCTIONS: EXPONENTIALS AND LOGARITHMIC GRAPH TRANSFORMATIONS
EXPRESSIONS AND FUNCTIONS

TRANSFORMATIONS OF THE EXPONENTIAL GRAPH

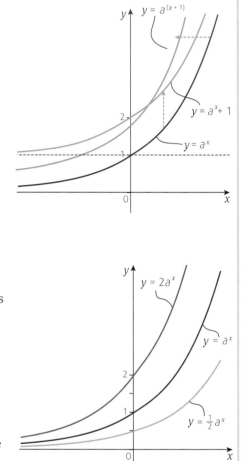

Exponential graphs can be transformed in the same way as other graphs. You may be given such a graph and asked to find its formula. We use the results from the section on transformations of graphs, (pages 24–25) applied to the graph of $y = a^x$.

Adding a constant to x moves the graph **horizontally**, and adding to the whole expression for y moves it **vertically**, so:

- $y = a^{(x+1)}$ is the graph of $y = a^x$ moved 1 unit **left** (blue graph).

- $y = a^x + 1$ is the graph of $y = a^x$ moved **up** 1 (green graph). A clue to this type would be the graph approaching the line $y = 1$ instead of the x-axis as $x \to -\infty$.

But notice that $y = a^{(x+1)} = a^x \times a^1 = a(a^x)$. The effect of this is to multiply all the y-values on the graph of $y = a^x$ by a constant, a, which means the result is the graph of $y = a^x$ stretched vertically.

Multiplying by k stretches the graph vertically if $k > 1$ and **compresses** it if $k < 1$.

Notice that a stretch transformation alters the distance between the asymptote line and the point of intersection of the graph with the y-axis, so look out for that.

$y = -a^x$ is the reflection of $y = a^x$ in the x-axis (so, a vertical-type transformation as you would expect).

$y = a^{-x}$ is the reflection in the y-axis.

contd

DON'T FORGET

If the alteration in the formula is to each appearance of x, then the alteration to the graph will be a horizontal one. If it is to the whole expression for y, the change to the graph will be vertical.

DON'T FORGET

Look out for hints for coordinates to substitute, for example where the graph cuts the y-axis so that $x = 0$, and also what line the graph gets closer and closer to.

TRANSFORMATIONS OF THE LOGARITHMIC GRAPH

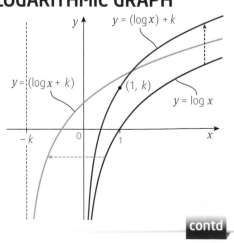

To be really on top of this topic, you need to be able to do the algebra of logs.

- $y = \log x + k$ will be the graph of $\log x$ **moved up k units** (down if k is negative) (pink graph).

- $y = \log (x + k)$ is $y = \log x$ **moved k to the left** (or right if $k < 0$) (green graph).

DON'T FORGET

A hint that the transformation is a $\log (x + k)$ type is that the graph will approach the line $x = -k$ instead of the line $x = 0$ (the y-axis).

Example: 1

Sketch the graph of $y = \log_3 81x$

Solution:

$\log_3 81x = \log_3 81 + \log_3 x = \log_3 3^4 + \log_3 x$
$= 4 \log_3 3 + \log_3 x = 4 + \log_3 x$

which means $y = \log_3 81x$ is the graph
of $y = \log_3 x$ moved up 4 units.

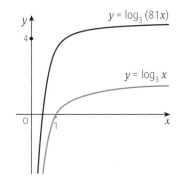

The graph of $y = k \log x$ will be the graph of $y = \log x$ stretched vertically by a factor of k.

Don't forget that, as $\log x^k = k \log x$, $y = \log x^k$ is just a different version of the same graph.

The graph of **$y = -\log x$** is the **reflection** of the graph of **$y = \log x$** in the x-axis:

$\log \frac{1}{x} = \log 1 - \log x = 0 - \log x$ so $y = \log \frac{1}{x}$ is the same graph as $y = -\log x$.

If you are given a graph to identify, you will almost certainly be given clues as to the type of transformation, such as the form of the answer; $y = \log_a (x - b)$, for example.

Remember that a basic log graph gets close to but doesn't touch the y-axis and also passes through the point $(1, 0)$. Look to see how these features have been moved – this will give you information.

You will also need to use the laws of logs.

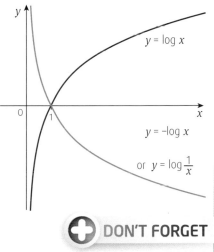

➕ **DON'T FORGET**

Substituting the coordinates of any points marked on the graph into the equation of the graph can be very useful.

Example: 2

The equation of the graph shown has formula $y = \log_b (x - a)$. Find a and b.

Solution:

The graph certainly looks like the basic log graph moved to the right 3 units, which makes $a = 3$.

This can be verified by substituting the coordinates of the point $(4, 0)$ into the formula:

$y = \log_b (4 - a) = 0$

and turning this into an exponential statement:

$b^0 = 4 - a$ (but $b^0 = 1$)

$a = 3$

Now, substituting the other point:

$\log_b (7 - a) = 2$

$\log_b 4 = 2$ (since $a = 3$)

And this gives $b^2 = 4$, so $b = 2$

The formula is $y = \log_2 (x - 3)$.

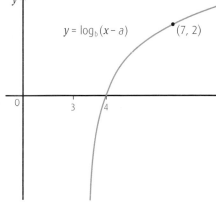

➡️ **ONLINE**

For more on surds and indices, follow the link at www.brightredbooks.net

💭 **THINGS TO DO AND THINK ABOUT**

1 Sketch the graph of $y = f(x)$, where $f(x) = 3 - \log_2(x + 4)$.   5

RECURRENCE RELATIONS 1  APPLICATIONS

Recurrence relations deal with situations where the **current value** depends on a **previous value**, and **future values** depend on the **current value**. Real-life examples of this include the depreciation in value of cars, removal of pollution by the use of a detergent or the actions of the sea, the spread of disease in a population, the growth of cells in a culture and the control of disease by the use of antibiotics.

FORMULAE FOR SEQUENCES

A recurrence relation is a rule, or formula, given for working out any term in a sequence if you know the one before.

For example, with u_n standing for the nth term in a sequence:

$u_{n+1} = u_n + 6$ (a recurrence relation) and $u_1 = -11$ (the first term)

generates the infinite sequence $-11, -5, 1, 7, 13, 19, 25, \dots$

The recurrence relation

is given by the formula $u_{n+1} = 2u_n + 3$. We could also use $u_n = 2u_{n-1} + 3$.

Evaluating terms

Given the formula and one term of a sequence, it is possible to find any others required.

Example: 1

Using the formula above, and given that $u_3 = 5$, find

(a) u_4, (b) u_6 and (c) u_2

Solution:

(a) $u_{n+1} = 2u_n + 3$

$u_4 = 2u_3 + 3$

$= 2 \times 5 + 3 = 13$

(b) Do this twice more to get to u_6 ... $u_5 = 29$, $u_6 = 61$

(c) Also from $u_{n+1} = 2u_n + 3$

we can write $u_3 = 2u_2 + 3$

$5 = 2u_2 + 3$

$u_2 = 1$

Example: 2

A sequence is defined by the linear recurrence relation $u_{n+1} = -4u_n + 7$

Express u_{n+2} in terms of u_n

Solution:

From the recurrence relation, we can write

$u_{n+2} = -4u_{n+1} + 7$

and then we can substitute for u_{n+1}, obtaining

$u_{n+2} = -4(-4u_n + 7) + 7$

and then, simplifying the expression,

$u_{n+2} = 16u_n - 21$

Finding the formula

With at least three terms of a particular recurrence relation, you can use simultaneous equations to find the formula $u_{n+1} = au_n + b$

Example: 3

Find the values of a and b in the sequence defined by $u_{n+1} = au_n + b$ where $u_3 = -1$, $u_4 = -9$ and $u_5 = -41$

Solution:

Using the three terms given, we can write:

$u_4 = au_3 + b$

and substitute to obtain $-9 = -a + b$

and $u_5 = au_4 + b$ giving $-41 = -9a + b$

Solve the pair of equations in a and b simultaneously:

Subtracting the second from the first gives $32 = 8a$ and so $a = 4$

Substituting 4 for a gives $b = -5$ and the recurrence relation is $u_{n+1} = 4u_n - 5$

LIMITS

Divergence

Sequences defined by recurrence relations of the form $u_{n+1} = au_n + b$ where $a < -1$ or $a > 1$ will continue to **decrease** or **increase forever**. They are said to **diverge**.

Such a sequence will have a graph like this:

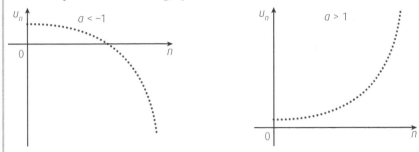

ONLINE

Check out the links at www.brightredbooks.net to find out more about recurrence relations and limits.

Convergence

Sequences defined by recurrence relations of the form $u_{n+1} = au_n + b$ where $-1 < a < 1$, **tend to a limit**. They approach a particular value and are said to **converge**.

Such a sequence will have a graph like this:

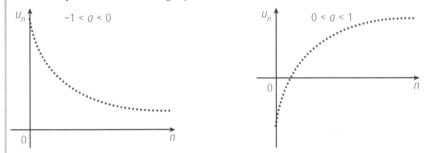

DON'T FORGET

A limit exists if $-1 < a < 1$

DON'T FORGET

When stating 'a limit exists since $-1 < a < 1$', a must have a value e.g. $-1 < 0.8 < 1$.

Any recurrence relation that has a limit will approach that limit in one of three ways. It may
- **oscillate above and below** the value of the limit until it finally settles on the limit
- **rise** to the limit
- **drop** to the limit

ONLINE TEST

Test yourself on your knowledge of this topic online at www.brightredbooks.net

In all cases as the recurrence relation approaches (tends to) the limit the values for u_n and u_{n+1} will become closer and closer until $u_{n+1} = u_n$.

Example: 4

Will the sequences given by the following recurrence relations converge?

$u_{n+1} = 3u_n - 7$ $u_{n+1} = 0.2u_n - 8$

Solution:

The sequence will not converge since a limit does not exist because $3 > 1$.

The sequence will converge since $-1 < 0.2 < 1$ and therefore a limit exists.

THINGS TO DO AND THINK ABOUT

1 A sequence is defined by $u_{n+1} = 3u_n + 4$ with $u_1 = 2$. What is the value of u_3?

A 34   B 21   C 18   D 13   2

RECURRENCE RELATIONS 2 APPLICATIONS

MORE ON LIMITS

Finding a limit

If a limit exists then both u_n and u_{n+1} will approach this limit.

Let the limit be L.

Then for $u_{n+1} = 0 \cdot 4u_n + 5$,

$$L = 0 \cdot 4L + 5$$
$$L - 0 \cdot 4L = 5$$
$$L(1 - 0 \cdot 4) = 5$$
$$L = \frac{5}{(1 - 0 \cdot 4)} = \frac{5}{0 \cdot 6} = \frac{50}{6} = \frac{25}{3}$$

Example: 1

Find the limit for the recurrence relation $u_{n+1} = \frac{2}{3}u_n + 10$

Solution:

A limit exists since $-1 < \frac{2}{3} < 1$. Let the limit be L.

$$L = \frac{2}{3}L + 10$$
$$\frac{1}{3}L = 10$$
$$L = 10 \div \frac{1}{3} = 30$$

Alternatively, since the general form for a recurrence relation is $u_{n+1} = au_n + b$, then when a limit exists

$L = aL + b$, so $L - aL = b$, $L(1 - a) = b$ which leads to $L = \frac{b}{(1 - a)}$

In this Example therefore, $L = \frac{10}{\left(1 - \frac{2}{3}\right)} = \frac{10}{\frac{1}{3}} = 30$

Example: 2

The sequence given by the recurrence relation $u_{n+1} = ku_n + 1 \cdot 2$ tends to a limit of 3 as $n \to \infty$. Determine the value of k.

Solution:

$L = kL + 1 \cdot 2$, but $L = 3$

so $3 = 3k + 1 \cdot 2$ which gives $3k = 1 \cdot 8$, hence $k = \frac{1 \cdot 8}{3} = 0 \cdot 6$.

Alternatively, using $L = \frac{b}{(1 - a)}$ with $L = 3$, $a = k$ and $b = 1 \cdot 2$ gives

$3 = \frac{1 \cdot 2}{(1 - k)}$, so $3(1 - k) = 1 \cdot 2$ which gives $3 - 3k = 1 \cdot 2$, $3k = 1 \cdot 8$, hence $k = \frac{1 \cdot 8}{3} = 0 \cdot 6$.

Example: 3

The following two recurrence relations generate sequences with the same limit.

(i) $P_{n+1} = tP_n + 4$ and (ii) $Q_{n+1} = t^2 Q_n + 2$

Determine the value of t and evaluate the limit.

Solution:

Find expressions for the limits of both formulae and equate them.

$$L_P = \frac{4}{(1 - t)}, \quad L_Q = \frac{2}{(1 - t^2)} \quad \text{so} \quad \frac{4}{(1 - t)} = \frac{2}{(1 - t^2)}$$
$$4(1 - t^2) = 2(1 - t)$$
$$2(1 - t^2) = 1 - t$$
$$2t^2 - t - 1 = 0$$

$(2t + 1)(t - 1) = 0, \Rightarrow t = -\frac{1}{2}, t = 1$

For a limit to exist, $-1 < t < 1, \Rightarrow t = -\frac{1}{2}$.

Evaluating the limit for this value, $L = \frac{4}{\left(1 - \left(-\frac{1}{2}\right)\right)} = \frac{4}{\frac{3}{2}} = \frac{8}{3}$. The limit is $\frac{8}{3}$.

PROBLEMS IN CONTEXT

Examples from real life include growth and decay, bank loans, pollution levels, inflation, policies for culling animals, removing pests and monitoring drug levels in patients. They normally involve changes over time. For example, a patient may be given a dose of antibiotics every four hours, or interest may be added to a loan every month. Use a '0' suffix for the amount at the start (A_0 rather than A_1) because then A_4 will mean the value after four of these time intervals. This avoids confusion!

Percentage increases and decreases often feature:

- 16% increase means that, after each interval, we have original 100% + 16% = 116% so × 1·16 to get new value.

- 9% decrease means that, after each interval, we have original 100% − 9% = 91% so × 0·91 to get new value.

Example: 1

(a) Pondweed is present in a pond and spreads during the summer growth period over a further $20\,m^2$. A team of volunteers clears 20% of the pondweed in an annual clean-up in the winter. If this volunteer work continues at the same rate, what area of the pond will be covered with pondweed each winter before the annual clean-up in the long term?

(b) The council decides that this amount of pondweed is still unsightly and decides to employ a team to clear the pond each winter so that the pondweed after each clean-up does not cover more than $40\,m^2$ of the pond. What percentage of the pondweed needs to be removed annually to achieve this target?

Solution:

(a) Before the annual clean-up:

$A_n = 0.8A_{n-1} + 20$ (20% removed; 80% remains, 20 m^2 extra during year).

Since $-1 < 0.8 < 1$, there will be a limit to the amount of pondweed in the long term.

Using $L = aL + b$ to find the limit, $L = 0.8L + 20$

$0.2L = 20$

$L = \frac{20}{0.2} = \frac{200}{2} = 100$

The pondweed will cover 100 m^2 in the long term.

(b) We need the value of a which gives the formula $A_n = aA_{n-1} + 20$ a limit of 40.

Substitute in the limit formula: $L = \frac{b}{1-a}$

$40 = \frac{20}{1-a}$

$40 - 40a = 20$

leading to $a = 0.5$.

This means that 50% must be removed (so, 50% remains) at each clean-up.

 DON'T FORGET

You must watch out for examples like this where the percentage in the question (20% here) is not the percentage you use for the calculations! (80%)

ONLINE

Check out the clip at www.brightredbooks.net to find out more about sequences and recurrence relations involved in problem solving such as the Towers of Hanoi.

THINGS TO DO AND THINK ABOUT

1. The terms of a sequence satisfy $u_{n+1} = ku_n + 5$. Find the value of k which produces a sequence with a limit of 4. **2**

2. A man decides to plant a number of fast-growing trees as a boundary between his property and the property of his next door neighbour. He has been warned, however, by the local garden centre that, during any year, the trees are expected to increase in height by 0·5 metres. In response to this warning he decides to trim 20% off the height of the trees at the start of any year.

   (a) If he adopts the 20% pruning policy, to what height will he expect the trees to grow in the long run? **3**

   (b) His neighbour is concerned that the trees are growing at an alarming rate and wants assurances that the trees will grow no taller than 2 metres. What is the minimum percentage that the trees will need to be trimmed each year so as to meet this condition? **3**

 **ONLINE TEST**

Test yourself on your knowledge of this topic online at www.brightredbooks.net

2 CALCULUS

DIFFERENTIATING FUNCTIONS
RELATIONSHIPS AND CALCULUS

DIFFERENTIATING AN ALGEBRAIC FUNCTION WHICH CAN BE SIMPLIFIED TO AN EXPRESSION IN POWERS OF x

Power rule

Rule: If $f(x) = ax^n$, $f'(x) = anx^{n-1}$

DON'T FORGET

Differentiation: "multiply by the index then reduce the index by one"

'Multiply the coefficient by the index, and reduce the index by 1.'

Here are some examples, differentiating with respect to x, except in Example 3 which differentiates with respect to t.

DON'T FORGET

Remember – the labels for the derivative need to match the question.

Example: 1

$y = 5x^3$

$\frac{dy}{dx} = 15x^2$

Example: 2

$f(x) = \frac{1}{2}x^9$

$f'(x) = \frac{9}{2}x^8$

Example: 3

$s = 6t^{\frac{1}{2}}$

$\frac{ds}{dt} = 3t^{-\frac{1}{2}}$

Since $a = ax^0$, the derivative of a, or any constant, is 0.

Negative indices are dealt with no differently from positive.

DON'T FORGET

Your basic arithmetic with fractions and negatives needs to be sound. Do lots of practice if necessary.

Example: 4

$f(x) = 4x^{-2}$

$f'(x) = -8x^{-3}$

Example: 5

$y = -x^{\frac{5}{6}}$

$\frac{dy}{dx} = \frac{5}{6}x^{-\frac{11}{6}}$

(since $-\frac{5}{6} - 1 = -\frac{5}{6} - \frac{6}{6}$

$= -\frac{11}{6}$)

Example: 6

$y = 5p + 8$

$\frac{dy}{dx} = 0$

(since there are no x-terms)

DON'T FORGET

'With respect to ...' tells you what to take as the variable for differentiating.

Differentiating a number of terms is no problem – just differentiate each term in order:

Example: 7

$y = 6x + 4x^3 - x^{-2} - 3x^{-\frac{3}{2}} + 5$

$\frac{dy}{dx} = 6 + 12x^2 + 2x^{-3} + \frac{9}{2}x^{-\frac{5}{2}}$

DON'T FORGET

You need to know all about indices and surds – practise a lot.

(In general, leave answers with fractions rather than decimals.)

Example 8 shows the roots rewritten as indices before differentiating.

Example: 8

Differentiate $y = 4\sqrt{x} + 5\sqrt[3]{x} + 7\sqrt[3]{x^2}$ with respect to x

Rewrite as

$y = 4x^{\frac{1}{2}} + 5x^{\frac{1}{3}} + 7x^{\frac{2}{3}}$

$\frac{dy}{dx} = 2x^{-\frac{1}{2}} + \frac{5}{3}x^{-\frac{2}{3}} + \frac{14}{3}x^{-\frac{1}{3}}$

Example 9 shows what to do when $f(x)$ involves brackets.

Example: 9

$f(x) = (x - 3)(x^2 + 5x - 8)$

$\quad\quad = x^3 + 2x^2 - 23x + 24$

$f'(x) = 3x^2 + 4x - 23$

contd

Example 10 shows how to deal with powers of x within fractional expressions. Take care when rewriting these expressions before differentiating:

Example: 10

$y = \frac{5}{x^3} = 5x^{-3}$

$\frac{dy}{dx} = -15x^{-4}$

Example: 11

$f(x) = \frac{54x^5}{6x^2} = 9x^3$

$f'(x) = 27x^2$

Example: 12

$f(x) = \frac{(x+4)(x-5)}{x^2}$

$= \frac{x^2 - x - 20}{x^2}$

$= \frac{x^2}{x^2} - \frac{x}{x^2} - \frac{20}{x^2}$

$= 1 - x^{-1} - 20x^{-2}$

The expression is in **differentiable** form, so it is ready to be differentiated. Remember, in spite of all this work so far, you still haven't differentiated!

$f'(x) = +x^{-2} + 40x^{-3} = \frac{1}{x^2} + \frac{40}{x^3}$

Example: 13

$\frac{d}{dx}\left(\frac{3x^5 + 2x^3 - 8x}{2\sqrt{x}}\right) = \frac{d}{dx}\left(\frac{3}{2}x^{\frac{9}{2}} + x^{\frac{5}{2}} - 4x^{\frac{1}{2}}\right)$

(remember, $\sqrt{x} = x^{\frac{1}{2}}$ and you must divide each term by $2x^{\frac{1}{2}}$)

$= \frac{27}{4}x^{\frac{7}{2}} + \frac{5}{2}x^{\frac{3}{2}} - 2x^{-\frac{1}{2}}$

Example: 14

$f(x) = 2\pi x^3$

π is a constant, so is treated like any other real number.

$f'(x) = 6\pi x^2$

 **DON'T FORGET**

In these last two examples, the expression is rewritten as a sum of terms – a string of separate terms in x added/subtracted together. This must always be done before differentiating.

EVALUATING DERIVATIVES

You could be asked to 'find $f'(3)$' or 'find the rate of change of f at $x = 3$'.

This involves substituting the value given into the expression for the derivative.

Example: 15

For the equation $y = 3x^3 + 2x - \frac{4}{x} + 3$, find the rate of change at $x = -1$.

$\frac{dy}{dx} = 9x^2 + 2 + 4x^{-2} = 9x^2 + 2 + \frac{4}{x^2}$

when $x = -1$, $\frac{dy}{dx} = 9 \times (-1)^2 + 2 + \frac{4}{(-1)^2}$

$= 9 + 2 + 4$

$= 15$

Example: 16

Find $f'(x)$ when $x = 4$ where

$f(x) = \frac{(x+4)(x-5)}{x^2}$ as given in Example 12.

$f'(x) = x^{-2} + 40x^{-3}$

$= \frac{1}{x^2} + \frac{40}{x^3}$

$f'(4) = \frac{1}{16} + \frac{40}{64}$

$= \frac{44}{64}$

$= \frac{11}{16}$

Note that 'Find $f'(4)$' would mean doing the same thing.

THINGS TO DO AND THINK ABOUT

1 If $y = 2x^{-3} + 3x^{\frac{4}{3}}$, $x > 0$, determine $\frac{dy}{dx}$.   2

2 The volume of a sphere is given by the formula $V = \frac{4}{3}\pi r^3$.
What is the rate of change of V with respect to r, at $r = 2$?   2

3 If $f(x) = \frac{1}{\sqrt[5]{x}}$, $x \neq 0$, what is $f'(x)$?   2

4 If $s(t) = t^2 - 5t + 8$, what is the rate of change of s with respect to t when $t = 3$?   2

5 What is the derivative of $\frac{1}{4x^3}$, $x \neq 0$?   2

6 $A = 2\pi r^2 + 6\pi r$. What is the rate of change of A with respect to r when $r = 2$?   2

ONLINE TEST

Head to www.brightredbooks.net and test yourself on differentiating functions.

DIFFERENTIATING USING THE CHAIN RULE
RELATIONSHIPS AND CALCULUS

THE CHAIN RULE

When dealing with a function of x (which will be written inside a bracket), raised to a power, the **chain rule** may be used.

Chain rule $\frac{d}{dx}[(f(x))^n] = n[f(x)]^{n-1} \times f'(x)$

Fortunately, actual examples don't look as complicated as this suggests!

Follow the working in the examples below to make sure you remember how to do it.

Example: 1
$$g(x) = (x + 6)^{-3}$$
$$g'(x) = -3(x + 6)^{-2}$$

Example: 2
$$y = \sqrt{x - 5}$$
$$= (x - 5)^{\frac{1}{2}}$$
$$\frac{dy}{dx} = \frac{1}{2}(x - 5)^{-\frac{1}{2}}$$
$$= \frac{1}{2\sqrt{x - 5}}$$

Example: 3
Differentiate $y = (3x + 5)^4$ with respect to x.

Solution:
Looking at the form it's in, we have $y = (f(x))^4$ contributing $4(f(x))^3$ to the answer.

We also need to include the derivative of $f(x)$, the expression in the bracket: $f'(x) = 3$

Putting it together gives
$$\frac{dy}{dx} = 4(3x + 5)^3 \times 3 = 12(3x + 5)^3$$

Example: 4
Find $\frac{dy}{dx}$ if $y = \sqrt{5x^2 + 3x - 7}$

Solution:
We need to rewrite: $y = (5x^2 + 3x - 7)^{\frac{1}{2}}$ to see that we have a similar form to the previous example.

Now use the chain rule
$$\frac{dy}{dx} = \frac{1}{2}(5x^2 + 3x - 7)^{-\frac{1}{2}} \times (10x + 3) = \frac{10x + 3}{2\sqrt{5x^2 + 3x - 7}}$$

DON'T FORGET

Try to leave your answers with positive indices and simplified as much as you can.

Example: 5
Differentiate $f(x) = \frac{7}{(2x - 3)^4}$

Solution:
First, rewrite as $f(x) = 7(2x - 3)^{-4}$
$$f'(x) = -28(2x - 3)^{-5} \times 2 = -56(2x - 3)^{-5} = \frac{-56}{(2x - 3)^5}$$

Example: 6
Differentiate $f(x) = (3 + \frac{1}{x})^5$

Solution:
$$f'(x) = 5(3 + \frac{1}{x})^4 \times -\frac{1}{x^2}$$
$$= -\frac{5}{x^2}(3 + \frac{1}{x})^4$$

Example: 7
Differentiate $f(x) = \frac{4}{(1 + \sqrt{x})} = 4(1 + x^{\frac{1}{2}})^{-1}$

Solution:
$$f'(x) = -1 \times 4(1 + x^{\frac{1}{2}})^{-2} \times \frac{1}{2}x^{-\frac{1}{2}}$$
$$= -\frac{-2}{\sqrt{x}(1 + \sqrt{x})^2}$$

ALTERNATIVE METHOD USING SUBSTITUTION

Chain rule: $\frac{dy}{dx} = \frac{dy}{du} \times \frac{du}{dx}$

To work Example 4 above this way, we would use substitution and write:

$u = 5x^2 + 3x - 7$ so that $y = (5x^2 + 3x - 7)^{\frac{1}{2}} = u^{\frac{1}{2}}$

$\frac{dy}{du} = \frac{1}{2}u^{-\frac{1}{2}}$

This will mean also that $\frac{du}{dx} = 10x + 3$

(differentiating u with respect to x)

Now we can substitute these expressions in $\frac{dy}{dx} = \frac{dy}{du} \times \frac{du}{dx}$

obtaining $\frac{dy}{dx} = \frac{1}{2}u^{-\frac{1}{2}} \times (10x + 3)$

but we need to substitute for u to complete it:

$= \frac{1}{2}(5x^2 + 3x - 7)^{-\frac{1}{2}}(10x + 3)$

as obtained earlier.

THINGS TO DO AND THINK ABOUT

1 What is the derivative of $(4 - 9x^4)^{\frac{1}{2}}$?

2 Given that $f(x) = (4 - 3x^2)^{-\frac{1}{2}}$ on a suitable domain, find $f'(x)$.

2

DON'T FORGET

Prepare functions for differentiating.

ONLINE

For more on differentiation, follow the link at www.brightredbooks.net

ONLINE

Head to www.brightredbooks.net/N5Maths and take the 'Surds and Indices' tests to check your prior knowledge and revise what you should know.

ONLINE

For more on surds and indices, follow the link at www.brightredbooks.net

DON'T FORGET

You don't need to know both ways. You can stick to the one you prefer.

ONLINE TEST

Want to test your knowledge of differentiation? Head to www.brightredbooks.net

DIFFERENTIATING: NATURE AND PROPERTIES OF FUNCTIONS 1

RELATIONSHIPS AND CALCULUS

THE GRADIENT OF A CURVE

The graph of a cubic equation $y = x^3 - 12x - 4$ is shown in the diagram.

Straight lines have constant gradients, but the gradient of a curve varies as you move along it. The gradient, $\frac{dy}{dx}$, gives the slope of the curve.

For $y = -3x + 5$, $\frac{dy}{dx} = -3$

So the slope of the derivative of $y = x^3 - 12x - 4$ is given by

$\frac{dy}{dx} = 3x^2 - 12$.

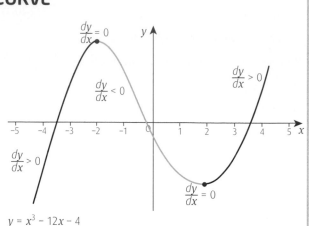

$$\frac{dy}{dx} = 0$$
$$\frac{dy}{dx} < 0$$
$$\frac{dy}{dx} > 0$$
$$\frac{dy}{dx} > 0$$
$$\frac{dy}{dx} = 0$$

$y = x^3 - 12x - 4$

DON'T FORGET

The question might use $f(x)$ and $f'(x)$ instead of y and $\frac{dy}{dx}$.

To find the gradient at any point on the curve, substitute the x-value of that point into the formula for $\frac{dy}{dx}$.

It can be seen from the diagram that the curve has zero gradient when $x = 2$ and $x = -2$.

This can be verified algebraically by calculating $\frac{dy}{dx}$. When $x = 2$, or -2, for example

$\frac{dy}{dx} = 3x^2 - 12 = 3 \times 2^2 - 12 = 12 - 12 = 0$

At the values $x = 2$ and -2, the **gradient = 0** and the function is neither increasing nor decreasing. It is said to be **stationary**.

DON'T FORGET

$\frac{dy}{dx} = 0$ for stationary points.

$\frac{dy}{dx} = 0$ **for stationary points.**

When $-2 < x < 2$

- the function is decreasing
- the gradient is negative
- the curve has a downward slope

$\frac{dy}{dx} < 0$

When $x < -2$ and $x > 2$

- the function is increasing
- the gradient is positive
- the curve has an upward slope

$\frac{dy}{dx} > 0$

TANGENT TO CURVES

Differentiation is used to determine the equation of a tangent to a curve at a given point.

The gradient of a curve at a given point is equal to the gradient of its tangent at the point of contact.

Two tangents to the curve $y = x^3 - 12x - 4$ have been drawn, at $x = -3$ and $x = 1$.

$x = -3 \therefore y = (-3)^3 - 12 \times (-3) - 4 = 5$

$\frac{dy}{dx} = 3x^2 - 12$

$\frac{dy}{dx} = 3 \times (-3)^2 - 12 = 15$

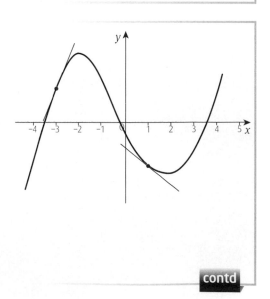

contd

so, at $(-3, 5)$ the gradient is 15.

Using $y - b = m(x - a)$, the equation of the tangent is $y - 5 = 15(x + 3)$ $\therefore$ $y = 15x + 50$

$x = 1$ $\therefore$ $y = 1^3 - 12 \times 1 - 4 = -15$

and $\frac{dy}{dx} = 3 \times 1^2 - 12 = -9$

so, at $(1, -15)$ the gradient is -9 and the equation of the tangent is $y = -9x - 6$.

Example: 1

Find the equation of the tangent at A, where $x = 4$, on the graph of $y = \frac{3}{\sqrt{x}}$, $x > 0$.

$y = \frac{3}{\sqrt{x}}$

A

Solution:

$y = 3x^{-\frac{1}{2}}$

$\frac{dy}{dx} = 3\left(-\frac{1}{2}x^{-\frac{3}{2}}\right) = \frac{-3}{2x^{\frac{3}{2}}} = \frac{-3}{2\sqrt{x^3}}$

At $x = 4$, $m = \frac{-3}{2\sqrt{4^3}} = \frac{-3}{16}$

and $y = \frac{3}{\sqrt{4}} = \frac{3}{2}$

Substituting these into

$y - b = m(x - a)$ and simplifying gives

$y - \frac{3}{2} = -\frac{3}{16}(x - 4)$, $16y - 24 = -3x + 12$ $\therefore$ $16y + 3x = 36$

Example: 2

The point Q lies on the curve with equation $y = x^3 - 3x^2$.

The tangent at Q has gradient -3. Find the equation of the tangent at Q.

Solution:

$\frac{dy}{dx} = 3x^2 - 6x = -3$

Write in standard quadratic form to give

$3x^2 - 6x + 3 = 0$

Factorise and solve for x: $3(x^2 - 2x + 1) = 0 \Rightarrow (x - 1)^2 = 0$

$\therefore$ $x = 1$

When $x = 1$, $y = -2$

substituting into $y - y_1 = m(x - x_1)$ gives $y = -3x + 1$

THINGS TO DO AND THINK ABOUT

1   A curve has equation $y = x^4 - 5x^3 + 6$.
    Find the equation of the tangent to this curve at the point where $x = -1$.   4

2   The point P $(3, -4)$ lies on the curve with the equation $y = x^2 - 5x + 2$.
    What is the gradient of the tangent to this curve at P?   2

3   What is the gradient of the tangent to the curve with the equation $y = x^3 - 4x + 3$ at
    the point where $x = 2$?   2

4   A curve has equation $y = 4x^5 - 17x$.
    What is the gradient of the tangent at the point $(1, -13)$?   2

DON'T FORGET

The tangent is a straight line with a constant gradient, touching the curve at the point where $x = a$,
$m_{tangent}$ at $a = \frac{dy}{dx}$ at a.

ONLINE

For more on gradients and equations of tangents follow the link at www.brightredbooks.net

ONLINE TEST

Want to test your knowledge of differentiation? Head to www.brightredbooks.net

DIFFERENTIATING: NATURE AND PROPERTIES OF FUNCTIONS 2

RELATIONSHIPS AND CALCULUS

INCREASING AND DECREASING FUNCTIONS

We can use $\frac{dy}{dx}$ to establish whether a function is **increasing**, **decreasing** or **stationary** at any point. The graph $y = x^3 - 12x - 4$ already discussed (shown on page 40) increases (gradient positive) and decreases (gradient negative), and it is stationary momentarily where it changes from one to the other.

Examples 1 and 2 show how to deal with questions about increasing and decreasing functions.

DON'T FORGET

'Always increasing' (or 'strictly increasing') isn't the same as 'never decreasing', as the first doesn't include the possibility of zero, whereas the second does.

Example: 1

Show that the function $y = \frac{1}{3}x^3 + 3x^2 + 11x + 4$ is always increasing.

DON'T FORGET

You need to know the method of completing the square.

Solution:

It's necessary to find $\frac{dy}{dx}$ and then show that it must be positive, whatever the value of x.

$$\frac{dy}{dx} = x^2 + 6x + 11$$
$$= (x^2 + 6x + 9) + 2$$
$$= (x + 3)^2 + 2 > 0 \text{ for all } x, \text{ so } y \text{ always increases.}$$

Example: 2

For what values of x is the graph of $y = 6 + 5x + 2x^2 - \frac{1}{3}x^3$ decreasing?

Solution:

We want $5 + 4x - x^2 < 0$.

Setting $\frac{dy}{dx} = 0$ and solving for x will tell you the values of x which make $\frac{dy}{dx}$ zero and then you can look at either side of these to find where $\frac{dy}{dx}$ is negative (function decreasing). A sketch often helps.

$$\frac{dy}{dx} = 5 + 4x - x^2 = 0 \qquad \therefore (5 - x)(1 + x) = 0 \qquad \therefore x = -1, 5$$

Sketching the parabola, $y = 5 + 4x - x^2$ ($-x^2$ so $\cap$-shaped) leads to this graph of the derivative. The graph shows when the gradient is < 0, and hence is negative, that is when the function is decreasing.

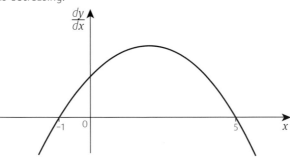

Alternatively, choosing a value of x between -1 and 5 and calculating $\frac{dy}{dx}$ at that point (e.g. $x = 0$ gives $\frac{dy}{dx} = 5$) will also give you the shape for this graph.

The graph is below the x-axis, $\frac{dy}{dx} < 0$, when x is less than -1 and when x is greater than 5, so we are able to say:

The function is decreasing when $x < -1$ and when $x > 5$.

THINGS TO DO AND THINK ABOUT

1 The graph of $y = f(x)$ shown has stationary points at $(0, p)$ and (q, r).

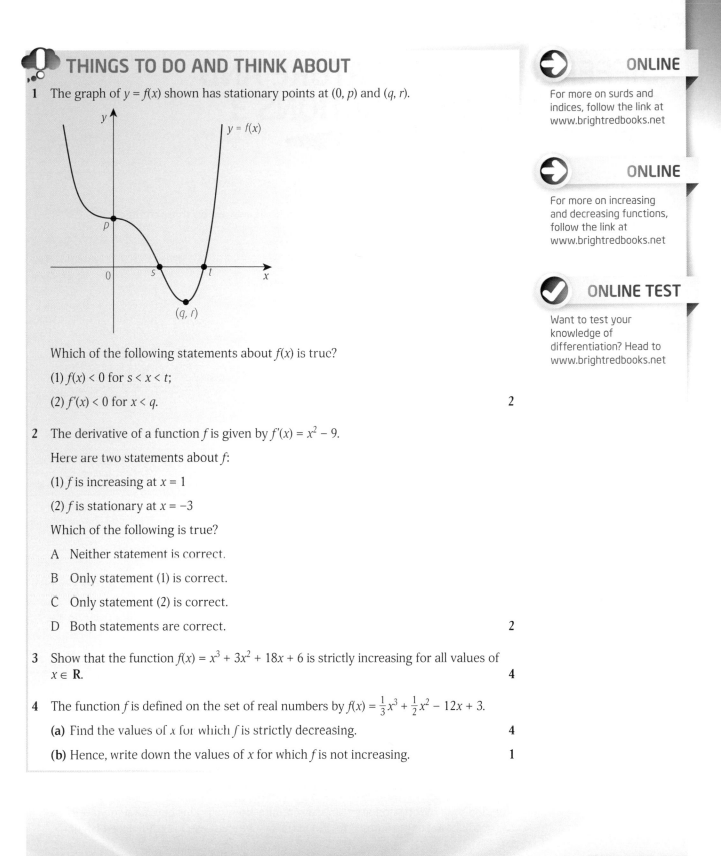

Which of the following statements about $f(x)$ is true?

(1) $f(x) < 0$ for $s < x < t$;

(2) $f'(x) < 0$ for $x < q$.   **2**

2 The derivative of a function f is given by $f'(x) = x^2 - 9$.

Here are two statements about f:

(1) f is increasing at $x = 1$

(2) f is stationary at $x = -3$

Which of the following is true?

A Neither statement is correct.

B Only statement (1) is correct.

C Only statement (2) is correct.

D Both statements are correct.   **2**

3 Show that the function $f(x) = x^3 + 3x^2 + 18x + 6$ is strictly increasing for all values of $x \in \mathbf{R}$.   **4**

4 The function f is defined on the set of real numbers by $f(x) = \frac{1}{3}x^3 + \frac{1}{2}x^2 - 12x + 3$.

(a) Find the values of x for which f is strictly decreasing.   **4**

(b) Hence, write down the values of x for which f is not increasing.   **1**

ONLINE

For more on surds and indices, follow the link at www.brightredbooks.net

ONLINE

For more on increasing and decreasing functions, follow the link at www.brightredbooks.net

ONLINE TEST

Want to test your knowledge of differentiation? Head to www.brightredbooks.net

DIFFERENTIATING: NATURE AND PROPERTIES OF FUNCTIONS 3
RELATIONSHIPS AND CALCULUS

CURVE-SKETCHING

In order to collect the information to sketch a curve, we do three things:

1 Find the stationary points and their nature.
   Solve the equation $\frac{dy}{dx} = 0$, use the 2nd differential or otherwise to investigate nature.

2 Find the points of intersection with the axes.
   To find where the graph intersects
   - the x-axis, set $y = 0$
   - the y-axis, set $x = 0$.

3 Investigate the behaviour of the curve for large positive and negative x.

DON'T FORGET

To get marks for your sketch, you must have evidence in your working.

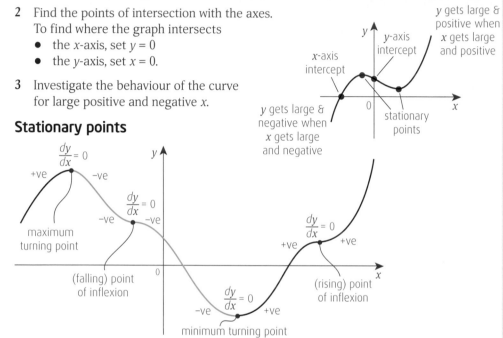

Stationary points

The points where the gradient of the tangent is zero are particularly important on the graphs you might have to sketch or interpret.

These are the **stationary** points of the curve, so the curve is neither rising nor falling at that point.

A stationary point can be
- a maximum or minimum turning point
- a point of inflexion.

A point of **inflexion** may occur on a rising curve (where the gradient is positive on either side) or on a falling curve (where the gradient is negative on either side). These are commonly called **rising** or **falling points of inflexion**.

Not all points of inflexion are stationary points, however. This one, for example, isn't – the gradient never becomes zero. But you won't study these until Advanced Higher!

Which of these options is the correct one depends on the gradient (value of $\frac{dy}{dx}$) on either side of the stationary point. In the diagram, signs for positive and negative gradients are shown on either side of the stationary points. Make sure you understand how the different combinations of positive and negative give the four possibilities.

contd

Maximum and minimum turning points can be determined by looking at the sign of f', $\left(\frac{dy}{dx}\right)$, near the critical point (**nature table**) or, alternatively, looking at the sign of f'', $\left(\frac{d^2y}{dx^2}\text{, the 2nd derivative}\right)$, at the critical point.

$f''(a) > 0 \implies$ minimum at $x = a$

$f''(a) < 0 \implies$ maximum at $x = a$

$f''(a) = 0 \implies$ stationary point or a point of inflexion.

Example: 1

Find the stationary points on the curve with equation $y = \frac{1}{4}x^4 - x^3$ and determine their nature.

Solution:

$\frac{dy}{dx} = x^3 - 3x^2 = 0$ for stationary points

$x^2(x - 3) = 0 \therefore x = 0$ or $x = 3$ when $x = 0, y = 0$; when $x = 3, y = \frac{1}{4}(3)^4 - (3)^3 = -6\frac{3}{4}$

$\frac{d^2y}{dx^2} = 3x^2 - 6x$:

when $x = 3, \frac{d^2y}{dx^2} = 3(3)^2 - 6(3) = 9 > 0 \implies$ a minimum turning point at $\left(3, -6\frac{3}{4}\right)$

when $x = 0, \frac{d^2y}{dx^2} = 3(0)^2 - 6(0) = 0$, so we look at the signs of $\frac{dy}{dx}$ either side of $x = 0$

x	-1	0	1
$\frac{dy}{dx}$	-4	0	-2
slope	\	—	\

the chosen value of $x > 0$ must not 'interfere' with the other stationary point at $x = 3$

Therefore there is a (falling) point of inflexion at $(0, 0)$.

If a graphic calculator is available, the stationary points could be checked by graphing the function.

Example: 2

Sketch the curve with equation $y = \frac{1}{4}x^4 - x^3$.

Solution:

In Example 1 above, we did step 1 and found the stationary points and their nature.

Now to step 2, finding where the curve intersects with the axes.

When $y = 0, \frac{1}{4}x^4 - x^3 = 0 \therefore x^3\left(\frac{1}{4}x - 1\right) = 0$, so $x = 0$ or $x = 4$, giving $(0, 0)$ and $(4, 0)$.

When $x = 0, y = \frac{1}{4}(0)^4 - (0)^3 = 0$, giving $(0, 0)$ which we already knew.

Finally step 3.

Only the largest power of x determines whether the curve is positive or negative at the extremities, so here we look at the sign of x^4.

Even powers will be positive, whether x is positive or negative, so,

as $x \to +\infty, y \to +\infty$

and as $x \to -\infty, y \to +\infty$

Putting all the information gathered onto a graph:

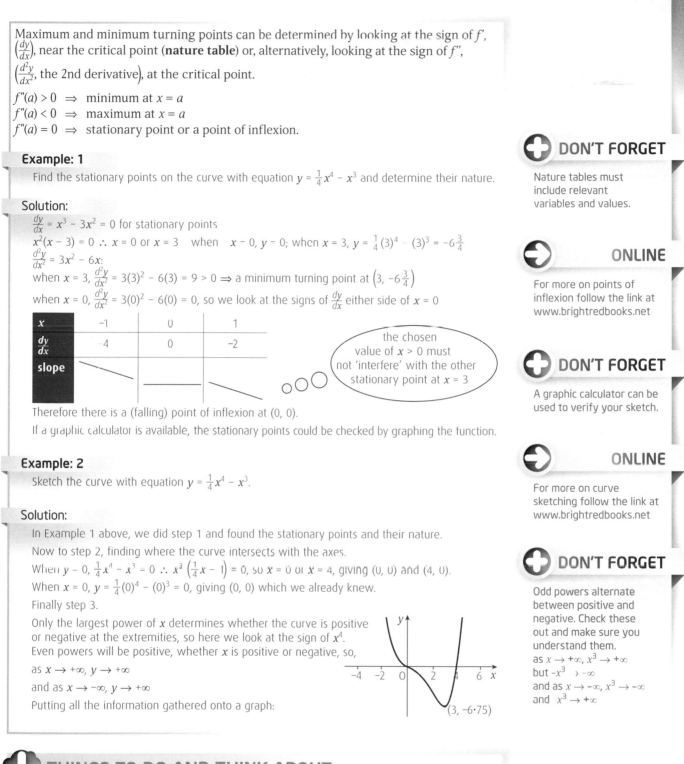

DON'T FORGET

Nature tables must include relevant variables and values.

ONLINE

For more on points of inflexion follow the link at www.brightredbooks.net

DON'T FORGET

A graphic calculator can be used to verify your sketch.

ONLINE

For more on curve sketching follow the link at www.brightredbooks.net

DON'T FORGET

Odd powers alternate between positive and negative. Check these out and make sure you understand them.

as $x \to +\infty, x^3 \to +\infty$

but $-x^3 \to -\infty$

and as $x \to -\infty, x^3 \to -\infty$

and $x^3 \to +\infty$

THINGS TO DO AND THINK ABOUT

1   A curve has equation $y = 3x^2 - x^3$.
   (a) Find the coordinates of the stationary points on this curve and determine their nature.   **6**
   (b) State the coordinates of the points where the curve meets the coordinate axes and sketch the curve.   **2**

2   A function f is defined on the set of real numbers by $f(x) = (x - 2)(x^2 + 1)$.
   (a) Find where the graph of $y = f(x)$ cuts
      (i) the x-axis   (ii) the y-axis.   **2**
   (b) Find the coordinates of the stationary points on the curve with equation $y = f(x)$ and determine their nature.   **8**

3   Find the coordinates of the turning points of the curve with equation $y = x^3 - 3x^2 - 9x + 12$ and determine their nature.   **8**

ONLINE TEST

Want to test your knowledge on the nature and properties of functions? Head to www.brightredbooks.net

INTEGRATING FUNCTIONS
RELATIONSHIPS AND CALCULUS

INTEGRATING – ANTI-DIFFERENTIATION

Apart from one complicating feature, if you integrate the result of a differentiation, you should get the original expression back. And vice versa: if you differentiate the result of an integration, you end up with the original expression.

The integral of $f(x)$ is written $\int f(x)dx$

Don't forget the 'dx'!

The diagram will remind you how to differentiate or integrate a term which is a power of x.

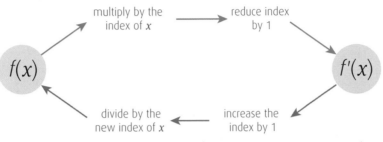

For example, to integrate the term $4x^3$, increase the index by 1, x^4,

divide coefficient by new index $\frac{4}{4} = 1$, so the answer is x^4

And, for this term, $\frac{6}{x^3}$, first rewrite as $6x^{-3}$,

then increase the index by 1, x^{-2},

divide coefficient by new index $\frac{6}{-2} = -3$, so the answer is $\frac{-3}{x^2}$.

Integration is slightly more complicated than differentiation because of the constant terms. The following three expressions are the same except for the constant terms:

$$\left.\begin{array}{l} 3x^2 - 5x + 4 \\ 3x^2 - 5x - 7 \\ 3x^2 - 5x \end{array}\right\} \text{ all differentiate to give } 6x - 5$$

Integrating $6x - 5$ gives the terms $3x^2 - 5x$ and it is impossible to know without further information what the constant term could be. We write '$+C$' to indicate the presence of a constant in the anti-derivative (which could of course be zero).

So, $\int(6x - 5)dx = 3x^2 - 5x + C$ and, unless there is information to enable you to find C (the constant of integration), that is how you should write the answer.

Integration rule

$\int kx^n dx = k\int x^n dx = \frac{kx^{n+1}}{n+1} + C$ provided $n \neq -1$

or $\frac{k}{n+1}x^{n+1} + C$

Example: 1

Integrate $\frac{7}{x^2}$ with respect to x.

Solution:

Rewrite $\frac{7}{x^2}$ as $7x^{-2}$ $\quad \int 7x^{-2}dx = \frac{7x^{-1}}{-1} + C = -\frac{7}{x} + C$

Example: 2

Find $\int p(2p + 1)(p - 6)dp$

Solution:

Brackets must be multiplied out first:
$$\int p(2p + 1)(p - 6)dp = \int(2p^3 - 11p^2 - 6p)dp$$
$$= \frac{2p^4}{4} - \frac{11p^3}{3} - \frac{6p^2}{2} + C$$
$$= \frac{1}{2}p^4 - \frac{11}{3}p^3 - 3p^2 + C$$

contd

Example: 3

Find $\int(3ps + qs^2)ds$

Solution:

'ds' tells us that s is the variable which we are integrating with respect to, and so p and q are constants (for the purposes of the integration, at least).

$\int(3ps + qs^2)ds = \frac{3ps^2}{2} + \frac{qs^3}{3} + C$

There is another important type of example which is all about checking that you understand that differentiation and integration are inverse processes:

Example: 4

a Differentiate $\sqrt{x^4 + 8}$

b Hence write down $\int\frac{x^3 dx}{\sqrt{x^4 + 8}}$

Solution:

Differentiating for part **a**:

$\frac{d}{dx}(x^4 + 8)^{\frac{1}{2}} = \frac{1}{2}(x^4 + 8)^{-\frac{1}{2}} \times 4x^3 = \frac{2x^3}{\sqrt{x^4 + 8}}$

Now part **b**:

Reversing the process, we can see that,

as $\int\frac{2x^3 dx}{\sqrt{x^4 + 8}} = \sqrt{x^4 + 8}$,

$\int\frac{x^3 dx}{\sqrt{x^4 + 8}} = \frac{1}{2}\sqrt{x^4 + 8} + C$

INTEGRATING USING THE CHAIN RULE

There is no chain rule equivalent for integration with composite functions, but the formula here may be useful:

Rule $\int(ax + b)^n dx = \frac{(ax + b)^{n+1}}{a(n + 1)} + C$ (provided $a \neq 0$, and $n \neq -1$)

ONLINE

For more on Integration follow the link at www.brightredbooks.net

Example: 5

Integrate $\frac{-60}{(5x - 2)^5}$ with respect to x.

Solution:

First, rewrite as $-60(5x - 2)^{-5}$

$\int 60(5x - 2)^{-5}dx = \frac{-60(5x - 2)^{-4}}{5 \times (-4)} + C = \frac{3}{(5x - 2)^4} + C$

Divide by new index and coefficient of x

ONLINE TEST

Want to test your knowledge on Integration? Head to www.brightredbooks.net

⚙ EXERCISE

Now try to integrate $\int\frac{-56}{(2x - 3)^5}dx$. When you have your answer, compare to Example 5 on p 38.

❗ THINGS TO DO AND THINK ABOUT

1 Find $\int(5x + 7)^2 dx$. 2

2 Find $\int x(4 - 3x)dx$. 2

3 Find $\int(1 - 8x)^{-\frac{1}{2}}dx$, where $x < \frac{1}{8}$. 2

4 Find $\int\left(\frac{1}{6x^2}\right)dx$, $x \neq 0$. 2

5 Find $\int(2x - 1)^{\frac{1}{2}}dx$, where $x > \frac{1}{2}$. 2

➕ DON'T FORGET

Integrate: increase index by 1, divide by new index.

DEFINITE INTEGRALS AND SOLVING DIFFERENTIAL EQUATIONS

RELATIONSHIPS AND CALCULUS

CALCULATING DEFINITE INTEGRALS

Using $F(x)$ to stand for the integral of $f(x)$,

$$\int_a^b f(x)dx = F(b) - F(a)$$

Method:

- **integrate** the expression
- **evaluate the result** for $x = b$ and for $x = a$, where a and b are the limits
- **find the difference**: **upper** limit value minus **lower** limit value.

Set out the working as shown here.

$\int_{-1}^{2} 5x^4 \, dx = [x^5]_{-1}^{2}$ integrate the expression

$= [2^5] - [(-1)^5]$ substitute the limit values and subtract

$= 32 + 1 = 33$

DON'T FORGET

Upper limit – lower limit.

DON'T FORGET

You are required to substitute zero, even when the result is zero, to demonstrate you know what you are doing.

DON'T FORGET

Show substitution of limits into expressions.

DON'T FORGET

First put in appropriate form to Integrate.

DON'T FORGET

It is usually easier to evaluate in 'surd' form.

Example: 1

Evaluate $\int_1^2 (4x^3 - x)dx$

Solution:

$\int_1^2 (4x^3 - x)dx$

$= \left[x^4 - \frac{1}{2}x^2\right]_1^2$

$= \left[(2)^4 - \frac{1}{2}(2)^2\right] - \left[(1)^4 - \frac{1}{2}(1)^2\right]$

$= [16 - 2] - \left[1 - \frac{1}{2}\right]$

$= 13\frac{1}{2}$

Example: 2

Evaluate $\int_1^3 (x^2 - 6x + 2)dx$

Solution:

$\int_1^3 (x^2 - 6x + 2)dx$

$= \left[\frac{1}{3}x^3 - \frac{6}{2}x^2 + 2x\right]_1^3$

$= \left[\frac{1}{3}(3)^3 - \frac{6}{2}(3)^2 + 2(3)\right] - \left[\frac{1}{3}(1)^3 - \frac{6}{2}(1)^2 + 2(1)\right]$

$= [9 - 27 + 6] - \left[\frac{1}{3} - 3 + 2\right]$

$= -11\frac{1}{3} \left(\text{or } -\frac{34}{3}\right)$

Example: 3

Evaluate $\int_4^9 (\sqrt{x} - 5)dx$

Solution:

$\int_4^9 (\sqrt{x} - 5)dx$

$= \int_4^9 \left(x^{\frac{1}{2}} - 5\right)dx$

$= \left[\frac{2}{3}x^{\frac{3}{2}} - 5x\right]_4^9$

$= \left[\frac{2}{3}(9)^{\frac{3}{2}} - 5(9)\right] - \left[\frac{2}{3}(4)^{\frac{3}{2}} - 5(4)\right]$

$= \left[\frac{54}{3} - 45\right] - \left[\frac{16}{3} - 20\right]$

$= -\frac{37}{3} \left(\text{or } -12\frac{1}{3}\right)$

Example: 4

Evaluate $\int_2^6 \frac{3}{\sqrt{4x+1}} dx$

Solution:

$\int_2^6 \frac{3}{\sqrt{4x+1}} dx$

$= \int_2^6 3(4x + 1)^{-\frac{1}{2}} dx$

$= \left[\frac{6}{4}(4x + 1)^{\frac{1}{2}}\right]_2^6$

$= \left[\frac{3}{2}\sqrt{4(6) + 1}\right] - \left[\frac{3}{2}\sqrt{4(2) + 1}\right]$

$= \frac{15}{2} - \frac{9}{2} = 3$

SOLVING DIFFERENTIAL EQUATIONS

We previously found the gradient formula of a curve from its equation. We can now reverse the process above to find the equation of a curve from its gradient formula:

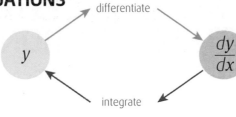

differentiate

$\dfrac{dy}{dx}$

integrate

Example: 5

Find the equation of the curve $y = f(x)$ for which $\dfrac{dy}{dx} = 2x + 5$ and which passes through the point $(0, 4)$.

Solution:

By integrating, we can find a family of curves which have this gradient formula:

$y = x^2 + 5x + C$ is the equation for this family.

Different values of C give a set of curves which appear parallel to each other.

Substituting the point $(0, 4)$ into the equation:

$y = x^2 + 5x + C$

we have $4 = 0^2 + 5 \times 0 + C \Rightarrow C = 4$

so, the parabola has equation $y = x^2 + 5x + 4$ (blue curve on diagram).

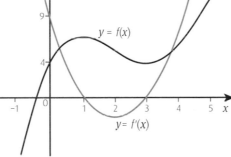

family of graphs
$y = x^2 + 5x + C$

Example: 6

The graph shows a cubic function $y = f(x)$ (in red) and its derived function $y = f'(x)$ (in green).

(a) Use the points of intersection of the graph of $f'(x)$ with the axes to find the equation of the derived function.

(b) Find the equation of $f(x)$.

Solution:

(a) The graph of $f'(x)$ cuts the x-axis at $x = 1$ and $x = 3$

$f'(x)$ has factors $(x - 1)$ and $(x - 3)$

$f'(x) = k(x - 1)(x - 3)$.

Expanding the brackets and substituting the point $(0, 9)$:

$f'(x) = 3x^2 - 12x + 9$ (check this for yourself).

(b) $f(x) = \int f'(x)dx = \int(3x^2 - 12x + 9)\,dx = x^3 - 6x^2 + 9x + C$

and, since $f(x)$ passes through $(0, 4)$, $C = 4$

$f(x) = x^3 - 6x^2 + 9x + 4$

THINGS TO DO AND THINK ABOUT

1 Given that $\int_4^t (3x + 4)^{-\frac{1}{2}}\, dx = 2$, find the value of t.

5

DIFFERENTIATING AND INTEGRATING TRIGONOMETRIC FUNCTIONS

DIFFERENTIATING AND INTEGRATING $\sin x$ AND $\cos x$

$$\frac{d}{dx}(\sin x) = \cos x \qquad \frac{d}{dx}(\cos x) = -\sin x$$

Differentiating the expression

$y = 3 \sin x + 4 \cos x - 5x$ gives $\frac{dy}{dx} = 3 \cos x - 4 \sin x - 5$

Since integrating is the inverse process, $\int \cos x \, dx = \sin x + C$
and $\int \sin x \, dx = -\cos x + C$

For example, $\int (3 \cos x + 5 \sin x + 3x^2) \, dx = 3 \sin x - 5 \cos x + x^3 + C$

The diagram might help you remember how to get the correct term.

Quite often we apply the chain rule when differentiating trigonometric functions.

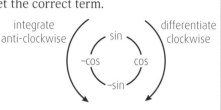

First, we can apply the chain rule in examples like sin( ) or cos( ) where the bracket contains a polynomial in x.

Example: 1

Differentiate $\cos 3x$ with respect to x.

Solution:

Differentiating cos gives $-\sin$, so we start with $-\sin 3x$

Multiplying this by the coefficient of $3x$, which is 3, gives the complete answer: $-3 \sin 3x$.

Example: 2

Differentiate $y = \sin(4x + \pi)$

Solution:

Differentiating sin gives cos, so we have $\cos(4x + \pi)$ and we multiply by the derivative of the expression in the bracket, which is 4, so

$\frac{dy}{dx} = 4 \cos(4x + \pi)$

Remember: $\sin(3x + 2)$ is different from $\sin 3x + 2$ or $\sin(3x) + 2$. In the first expression, the angle is $(3x + 2)$, but in the second and third, the angle is $3x$ and then 2 is added to the sine function.

We can also apply the chain rule to functions which are powers of $(\sin x)$ or $(\cos x)$ rather than powers of x.

Example: 3

Differentiate $y = \sin^4 x = (\sin x)^4$

Solution:

It is sometimes clearer to rewrite powers of trig functions like this.

$\frac{d}{dx}[(f(x))^n] = n[f(x)]^{n-1} \times f'(x)$

Here $f(x) = \sin x$ so $f'(x) = \cos x$ and $n = 4$

$\frac{dy}{dx} = 4(\sin x)^3 \times \cos x = 4 \sin^3 x \cos x$

4 to the front, bracket to one lower power, multiply by differential of bracket

contd

Example: 4

Differentiate $y = \sin\sqrt{x}$

Solution:

Method 1

$\frac{dy}{dx} = \cos\sqrt{x} \times \frac{d}{dx}\left(x^{\frac{1}{2}}\right)$

$\quad = \cos\sqrt{x} \times \frac{1}{2}x^{-\frac{1}{2}}$

$\quad = \frac{\cos\sqrt{x}}{2\sqrt{x}}$

Method 2

Using substitution

$\frac{dy}{dx} = \frac{dy}{du} \times \frac{du}{dx}$

Let $u = \sqrt{x} = x^{\frac{1}{2}}$ $\quad$ so $\frac{du}{dx} = \frac{1}{2}x^{-\frac{1}{2}}$

$\quad y = \sin u$ $\quad$ so $\frac{dy}{du} = \cos u$

$\frac{dy}{dx} = \frac{dy}{du} \times \frac{du}{dx} = \cos u \times \frac{1}{2}x^{-\frac{1}{2}} = \cos\sqrt{x} \times \frac{1}{2}x^{-\frac{1}{2}}$

$\quad = \frac{\cos\sqrt{x}}{2\sqrt{x}}$

Example: 5

Find $\frac{dy}{dx}$ when $y = (3\cos x - 2)^4$

Solution:

Using the substitution method:

Let $u = 3\cos x - 2$ so that

$\frac{du}{dx} = -3\sin x$ and $y = u^4$

$\frac{du}{dx} = \frac{dy}{du} \times \frac{du}{dx}$

$\quad = 4u^3 \times -3\sin x$

$\quad = 4(3\cos x - 2)^3 \times -3\sin x$

$\quad = -12\sin x\,(3\cos x - 2)^3$

Alternatively, $y = (3\cos x - 2)^4$

$\frac{du}{dx} = 4(3\cos x - 2)^3 \times \frac{d}{dx}(3\cos x - 2)$

$\quad = 4(3\cos x - 2)^3\,(-3\sin x)$

$\quad = -12\sin x\,(3\cos x - 2)^3$

DON'T FORGET

Angles must always be in radians for this topic – not degrees.

INTEGRATING $\sin(ax + b)$ AND $\cos(ax + b)$

Working backwards from differentiation, we have:

$\int \sin(ax + b)\,dx = -\frac{1}{a}\cos(ax + b) + C$

$\int \cos(ax + b)\,dx = \frac{1}{a}\sin(ax + b) + C$

we divide by the coefficient of x

You can differentiate the right-hand sides to check that you obtain the left-hand sides.

Example: 6

Integrate $\cos(3x - 1)$ with respect to x.

Solution:

$\int \cos x$ gives $\sin x$ and dividing by the coefficient of $3x - 1$, we get the answer

$\frac{1}{3}\sin(3x - 1) + C$

Example: 7

Find $\int \sin(\pi - 2x)\,dx$

Solution:

$\int \sin(\pi - 2x)\,dx = -\cos(\pi - 2x) \div -2 + C$

$\quad = \frac{1}{2}\cos(\pi - 2x) + C$

DON'T FORGET

Some rules for integration are given in the formula list at the beginning of the exam paper.

DON'T FORGET

Radians must be used.

Definite integrals

Calculating definite integrals of functions of the form $f(x) = p\cos(qx + r)$ and $f(x) = p\sin(qx + r)$.

DON'T FORGET

Learn exact values.

Example: 8

Evaluate $\int_0^{\frac{\pi}{2}} 2\sin 3x\,dx$

Solution:

$\int_0^{\frac{\pi}{2}} 2\sin 3x\,dx = \left[-\frac{2}{3}\cos 3x\right]_0^{\frac{\pi}{2}}$

$\quad = \left[-\frac{2}{3}\cos 3\left(\frac{\pi}{2}\right)\right] - \left[-\frac{2}{3}\cos 3(0)\right]$

$\quad = [0] - \left[-\frac{2}{3}(1)\right]$

$\quad = \frac{2}{3}$

Example: 9

Find the exact value of $\int_{-\frac{\pi}{3}}^{\frac{\pi}{3}}(1 - \cos 2x)\,dx$.

Solution:

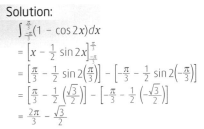

$\int_{-\frac{\pi}{3}}^{\frac{\pi}{3}}(1 - \cos 2x)\,dx$

$\quad = \left[x - \frac{1}{2}\sin 2x\right]_{-\frac{\pi}{3}}^{\frac{\pi}{3}}$

$\quad = \left[\frac{\pi}{3} - \frac{1}{2}\sin 2\left(\frac{\pi}{3}\right)\right] - \left[-\frac{\pi}{3} - \frac{1}{2}\sin 2\left(-\frac{\pi}{3}\right)\right]$

$\quad = \left[\frac{\pi}{3} - \frac{1}{2}\left(\frac{\sqrt{3}}{2}\right)\right] - \left[-\frac{\pi}{3} - \frac{1}{2}\left(-\frac{\sqrt{3}}{2}\right)\right]$

$\quad = \frac{2\pi}{3} - \frac{\sqrt{3}}{2}$

DON'T FORGET

To include '+ C' when integrating.

ONLINE TEST

Head to www.brightredbooks.net to test yourself on differentiating and integrating trigonometric functions.

THINGS TO DO AND THINK ABOUT

1   Given that $y = \sin(x^2 - 3)$, find $\frac{dy}{dx}$.   **2**

2   If $y = 3\cos^4 x$, find $\frac{dy}{dx}$.   **2**

3   Given that $f(x) = 4\sin 3x$, find $f'(0)$.   **2**

4   Find $\int(2x^{-4} + \cos 5x)\,dx$.   **2**

DIFFERENTIATION: MATHEMATICAL MODELLING AND PROBLEM SOLVING APPLICATIONS

GREATEST AND LEAST VALUES ON A CLOSED INTERVAL

To find the maximum and/or minimum values of a function on a closed interval we consider the 'end points' and the turning points.

Example: 1

What are the maximum and minimum values of $y = x^3 - 12x - 4$ within the interval $0 \leq x \leq 4$?

Solution:

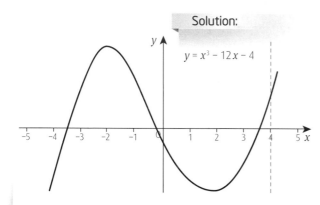

$y = x^3 - 12x - 4$

We look for the highest and lowest value of y between the y-axis and the line $x = 4$, including on those lines. The maximum and minimum values for a continuous function will always occur at the end points or at the turning points.

We know that the function $y = x^3 - 12x - 4$ has a turning point at $x = 2$. (see p 40 for a reminder).

At this point $y = (2)^3 - 12(2) - 4 = -20$.

At the end points: $x = 0$, $y = (0)^3 - 12(0) - 4 = -4$
$x = 4$, $y = (4)^3 - 12(4) - 4 = 12$

Hence, the maximum value is 12 (when $x = 4$) and the minimum value is -20 (when $x = 2$).

DON'T FORGET ➕

You must state that the derivative is zero for the maximum and/or minimum values.

DON'T FORGET ➕

Marks will be lost if the nature of the stationary value is not verified by using the second derivative or a nature table.

OPTIMISATION

Optimisation is the process of finding the best/largest/highest/loudest of various options – or maybe the lowest/cheapest/smallest option. It is **optimising the possibility**. We are effectively finding the **maxima** or **minima**.

We use the knowledge that maximum or minimum values of a function $f(x)$ occur when $f'(x) = 0$, or at a boundary of a closed interval.

Example: 2

To advertise a performance of an Edinburgh Festival Fringe show, flyers are to be distributed.

The profit made from the sale of tickets, £P, sold for the performance is related to the number of flyers, n hundred, distributed by this formula: $P = 12n^3 - n^4$.

What number of flyers maximises the profit made, and how much profit is that?

Solution:

Maximum and minimum values occur when the derivative is zero, so find $\frac{dP}{dn}$ and solve $\frac{dP}{dn} = 0$.

$\frac{dP}{dn} = 36n^2 - 4n^3 = 0 \Rightarrow 4n^2(9 - n) = 0$ so $n = 0$ or $n = 9$

It may seem obvious that n will be 9, giving 900 flyers distributed, rather than zero, but this needs to be confirmed.

$\frac{d^2P}{dn^2} = 72n - 12n^2$, when $n = 0$, $\frac{d^2P}{dn^2} = 0$ and $P = 0$.

When $n = 9$, $\frac{d^2P}{dn^2} = 72(9) - 12(9)^2 = -324 < 0 \Rightarrow$ a maximum.

Alternatively, evaluate $\frac{dP}{dn}$ on either side of $n = 9$ and display in a nature table.

n	8	9	10
$\frac{dP}{dn}$	256	0	−400
$\frac{dP}{dn}$	+ve	0	−ve
shape	╱	—	╲

So, when $n = 9$, P is a maximum.

When $n = 9$, $P = 12(9)^3 - (9)^4 = 2187$.

A maximum profit of £2187 is made when 900 flyers are distributed.

contd

Mathematical modelling

In examination questions on optimisation, you are often asked to show that the mathematical model given is correct in the first part of the question, then asked to maximise or minimise in the second.

In optimisation questions, because the model will usually be given, you should be able to do the second part even if you cannot do the first part.

DON'T FORGET

For optimisation questions, try the 2nd part even if the 1st part seems impossible.

Example: 3

A rectangular plot of land is to be fenced off against a straight wall using 12 metres of fencing.
(a) Find an expression for the area of the plot. (This is setting up the mathematical model.)
(b) Find the dimensions of the largest rectangle which can be fenced off. (This is the optimisation.)

Solution:

(a) Side parallel to the wall will measure 12 – 2x metres
Area of rectangle, $A = x(12 - 2x) = 12x - 2x^2$

(b) $\frac{dA}{dx} = 12 - 4x = 0$ for a maximum/minimum area, therefore $x = 3$.
$\frac{d^2A}{dx^2} = -4 < 0 \Rightarrow$ a maximum.
So, a rectangle with dimensions 3m and 6m will give the largest area.

wall

x m

$(12 - 2x)$m

Example: 4

A window is to be made in the shape of a rectangle with a semicircular section above, and must have a perimeter of 8 metres.
(a) Given that the width of the window is $2x$, show that the area, A, of the glass is given by the formula $A = 8x - 2x^2 - \frac{1}{2}\pi x^2$.
(b) The designer wants to maximise the area in order to let in the maximum amount of light. Find the width of the window which will do this. Give your answer to the nearest cm.

Solution:

(a) Let the height of the rectangular section be h, then the total length of the three straight edges and the semicircle is $2x + 2h + \pi x$.

$P = 2x + 2h + \pi x = 8$, so $2h = 8 - 2x - \pi x$

Area = area of rectangle + area of semicircle
$A = 2xh + \frac{1}{2}\pi x^2$
$= x(8 - 2x - \pi x) + \frac{1}{2}\pi x^2$
$= 8x - 2x^2 - \pi x^2 + \frac{1}{2}\pi x^2$
$= 8x - 2x^2 - \frac{1}{2}\pi x^2$

(b) Max/min values of area will occur when
$\frac{dA}{dx} = 0$
$\frac{dA}{dx} = 8 - 4x - \pi x = 0$
$(4 + \pi)x = 8 \Rightarrow x = \frac{8}{4 + \pi} = 1\cdot1202$ metre
$\frac{d^2A}{dx^2} = -4 - \pi < 0 \Rightarrow$ a maximum. Hence the width of window which gives the maximum area is 2·24 metres (to the nearest cm).
Alternatively, evaluate $\frac{dA}{dx}$ on either side of $x = 1\cdot12$ and display in a nature table.

h

$2x$

ONLINE

For more on optimisation, follow the link at www.brightredbooks.net

ONLINE TEST

Head to www.brightredbooks.net to test yourself on this topic.

THINGS TO DO AND THINK ABOUT

1 A manufacturer is asked to design an open-ended shelter, as shown, subject to the following conditions.

   1 The frame of a shelter is to be made of rods of two different lengths:
      • x metres for top and bottom edges; • y metres for each sloping edge.

   2 The frame is to be covered by a rectangular sheet of material. The total area of the sheet is 24 m².

y

x

(a) Show that the total length, L metres, of the rods used in a shelter is given by
$L = 3x + \frac{48}{x}$. 3

(b) The rods cost £8·25 per metre. To minimise production costs, the total length of rods used for a frame should be as small as possible.
(i) Find the value of x for which L is a minimum.
(ii) Calculate the minimum cost of the frame. 7

INTEGRATION: MATHEMATICAL MODELLING AND PROBLEM SOLVING ⬭APPLICATIONS⬭

AREAS

Finding the area between a curve and the x-axis

When asked to calculate the area **under a curve** we are calculating the area between the curve, $y = f(x)$, and the x-axis, $y = 0$.

$$\text{Area} = \int (f(x) - 0)dx$$

When we are asked to calculate the area **under the x-axis** we are calculating the area between the x-axis and the curve.

$$\text{Area} = \int (0 - f(x))dx$$

In both cases area is $= \int(\text{upper} - \text{lower})dx$.

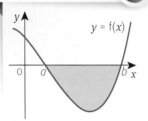
Example: 1

Find the area enclosed between the curve with equation $y = x^3 - 3x^2 - x + 3$ and the x-axis, as shown in the diagram.

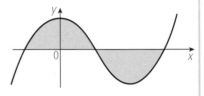

Solution:

We need points of intersection of the curve and the x-axis. Factorising the equation of the curve, we obtain $y = (x + 1)(x - 1)(x - 3)$ and setting $y = 0$ (x-axis), we obtain the roots $x = -1, 1, 3$.

The area above the x-axis is given by

$\int_{-1}^{1}(x^3 - 3x^2 - x + 3)dx$ equation of curve (although this is $\int(f(x) - 0)dx$, we do not need to write −0).

Limits obtained from the intersection points.

$= \left[\frac{1}{4}x^4 - x^3 - \frac{1}{2}x^2 + 3x\right]_{-1}^{1}$ integrating

$= \left[\frac{1}{4}(1)^4 - (1)^3 - \frac{1}{2}(1)^2 + 3(1)\right] - \left[\frac{1}{4}(-1)^4 - (-1)^3 - \frac{1}{2}(-1)^2 + 3(-1)\right]$ substituting limits

$= \left[\frac{1}{4} - 1 - \frac{1}{2} + 3\right] - \left[\frac{1}{4} + 1 - \frac{1}{2} - 3\right] = 4$ take care when evaluating with negatives

The area below the x-axis is given by

$-\int_{1}^{3}(x^3 - 3x^2 - x + 3)dx$ or $\int_{3}^{1}(x^3 - 3x^2 - x + 3)dx$, reversing the limits has the same effect.

Evaluating this integral gives 4.

Therefore the total area is $4 + 4 = 8$ square units.

Finding the area between a straight line and a curve or two curves

To find an area, you need
- the formulae for both curves
- the x-coordinates of the points where the curves intersect, to give the limits a and b.

$$\text{Area} = \int_a^b (f(x) - g(x))dx$$

Example: 2

Calculate area A on the graph. $f(x) = x^3 - 20x + 25$

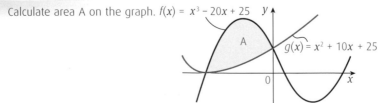

contd

Solution:

Solve $x^3 - 20x + 25 = x^2 + 10x + 25$ to find intersections

$\Rightarrow x^3 - x^2 - 30x = 0 \Rightarrow x(x^2 - x - 30) = 0 \Rightarrow x(x - 6)(x + 5) = 0$

$\Rightarrow x = 0,\ x = 6,\ x = -5$

From the sketch in the question, we can see that -5 and 0 are the values we will need to evaluate between.

$\text{Area} = \int_{-5}^{0} ((x^3 - 20x + 25) - (x^2 + 10x + 25))dx$

$= \int_{-5}^{0} (x^3 - x^2 - 30x)dx$

$= \left[\frac{x^4}{4} - \frac{x^3}{3} - 15x^2\right]_{-5}^{0}$

$= 177$ (3 sf)

In Example 2, if no sketch had been given, it would be a good idea to draw one. To decide which is the upper curve, take an x-value between the intersection points and substitute into both equations for the curves. The one which gives the higher y-value will be the equation of the upper curve.

Example: 3

Find the area enclosed by the graph $y = \cos 2x$, the x-axis and the lines $x = 0$ and $x = 2\pi$

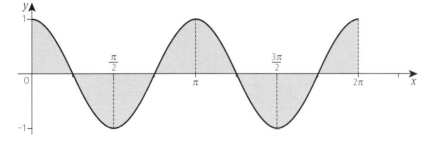

Solution:

The area is partly above and partly below the x-axis. From the symmetry, we know that the areas above and below are equal. However, as the area for each section of $\frac{\pi}{4}$ radians has identical area (look at the graph), an efficient way of working would be to calculate

$8\int_{0}^{\frac{\pi}{4}} \cos 2x\,dx = 8\left[\frac{1}{2}\sin 2x\right]_{0}^{\frac{\pi}{4}} = 4\sin\frac{\pi}{2} - 4\sin 0 = 4 - 0 = 4$

ONLINE

For more on areas, follow the link at www.brightredbooks.net

THINGS TO DO AND THINK ABOUT

1 The diagram shows the curve with equation $y = x^3 - x^2 - 4x + 4$ and the line with equation $y = 2x + 4$. The curve and the line intersect at the points $(-2, 0)$, $(0, 4)$ and $(3, 10)$. Find the shaded area.

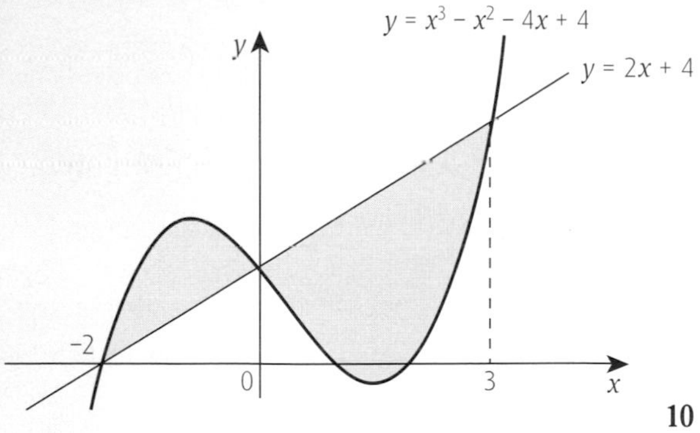

10

2 The diagram shows graphs with equations $y = 14 - x^2$ and $y = 2x^2 + 2$. Write down and simplify an integral that represents the shaded area.

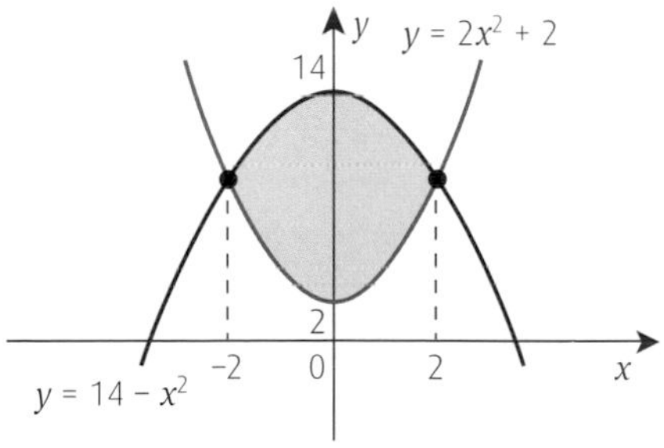

3

ONLINE TEST

Head to www.brightredbooks.net to test yourself on this topic.

RATES OF CHANGE: MATHEMATICAL MODELLING AND PROBLEM SOLVING APPLICATIONS

USING THE RATE OF CHANGE

$f'(x)$ or $\frac{dy}{dx}$ measures the **rate of change**, but so far we have used rate of change in the context of gradients and curves and discussed the situation from a geometric point of view.

There are other contexts where the same processes of calculus are used.

Example: 1

Find the rate of change of the function $g(x) = 2x^2 - \sqrt{x}$ when $x = 3$.

Solution:

The rate of change is $g'(x)$. If $x = 3$, we need to calculate $g'(3)$.

$g(x) = 2x^2 - x^{\frac{1}{2}}$

$g'(x) = 4x - \frac{1}{2}x^{-\frac{1}{2}}$ and so $g'(3) = 12 - \frac{1}{2\sqrt{3}}$

Example: 2

A balloon is attached to a pump and is expanding so that its diameter in cm after t seconds is given by the formula $d(t) = 18\sqrt[3]{t}$. What is the rate of change of the diameter after 8 seconds?

Solution:

$d(t) = 18t^{\frac{1}{3}}$ Rate of change is $d'(t) = 6t^{-\frac{2}{3}}$

For the rate of change after 8 seconds, calculate $d'(8)$:

$d'(8) = \frac{6}{\sqrt[3]{64}} = \frac{6}{4} = 1.5$ cm per second

That's the answer, but of course you can solve problems much better if you know what they're all about. The rate of change tells us that the balloon's diameter is increasing with respect to t at 1·5 cm/s after 8 seconds, but this is only true for a moment. Common sense tells us that the diameter increases fast when the balloon starts being blown up and more slowly as time passes. After 8 seconds, the balloon's diameter is 36 cm, and increasing slowly.

Example: 3

The current in an electrical circuit is given by the formula $I(R) = \frac{250}{R}$ amps where R is the resistance measured in ohms. When the resistance is 15 ohms, what is the rate of change of the current, I?

Solution:

$I(R) = 250R^{-1}$

$\Rightarrow I'(R) = -250R^{-2}$

$R = 15$ so calculate $I'(15)$

$= -250 \times \frac{1}{15^2} = -\frac{250}{225} = -1.11$ amps/ohm

Example: 4

This example demonstrates the use of trigonometry when calculating rates of change.

The water level in a reservoir during a month of drought varies according to the formula

$D = 3\cos(0.2t) + 4$

where D is the depth of water in metres at noon and t is the number of days since the drought began.

(a) Find the water level at the start of the drought.

(b) What is the water level at noon on day 6?

(c) What is the rate of change at noon on day 10?

Solution:

(a) $D = 3\cos(0) + 4 = 7$ metres

(b) $D = 3\cos(0.2 \times 6) + 4 = 5.1$ metres

(c) $D'(t) = -3\sin(0.2t) \times 0.2 = -0.6\sin(0.2t)$
$D'(10) = -0.6\sin(2) = -0.5$ metres (to 1 d.p.).

(This means that, at noon on day 10, the water level is falling at a rate of half a metre per day, although it won't actually fall by half a metre that day because the rate is constantly changing.)

contd

SPEED, DISTANCE, TIME AND ACCELERATION

Distance (or displacement), speed (velocity) and acceleration can all be represented as functions of **time**.

The symbol used for distance is s.

Speed (velocity) is represented by the symbol v and is the rate of change of distance, s, with time.

$$v = \frac{ds}{dt}$$

Acceleration is the rate of change of velocity with time.

$$a = \frac{dv}{dt}$$

Example: 5

The distance travelled in metres is given by the formula
$s = 4t + 8t^2$ (t is the time in seconds from the start).
What is the speed 6 seconds after setting off?

Solution:

The speed, v, is given by $\frac{ds}{dt}$

$\frac{ds}{dt} = 4 + 16t$
and when $t = 6$, $\frac{ds}{dt} = 4 + 96 = 100\,\text{m/s}$

Example: 6

A projectile travels at a speed v m/s where $v = 12t - t^2$. How far does the projectile go in the first 10 seconds after launch, to the nearest metre?

Solution:

$v = \frac{ds}{dt}$ and so
$s = \int v\, dt = \int (12t - t^2)\, dt = 6t^2 - \frac{1}{3}t^3 + C$

The projectile had travelled zero distance at time zero, so $(0, 0)$ must fit the equation.

Hence $C = 0$.

Distance $= 6t^2 - \frac{1}{3}t^3 = 6 \times 100 - \frac{1000}{3} = 267\,\text{m}$

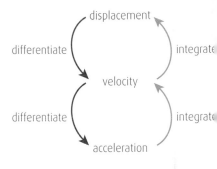

DIFFERENTIAL EQUATIONS

From a given rate of change and initial conditions, we can determine and use a function. This is the start to solving differential equations.

Example: 7

The rate at which water is emptying from a tank varies according to $V'(t) = -\frac{4}{\sqrt{t}}$, $0 < t \le 6$, where t is the time in hours and V is the volume of water in cubic metres.
The tank contains $20\,\text{m}^3$ of water initially. Find the volume of water remaining after four hours.

Solution:

$V'(t) = -4t^{-\frac{1}{2}}$ first write in integrating form

$V(t) = -\frac{4t^{\frac{1}{2}}}{\frac{1}{2}} + C = -8\sqrt{t} + C$ integrate

$t = 0$, $V = 20 \Rightarrow V(0) = 8\sqrt{0} + C = 20 \rightarrow C = 20$

$\therefore V(t) = 20 - 8\sqrt{t}$

$V(4) = 20 - 8\sqrt{4} = 4$. So $4\,\text{m}^3$ of water remains after four hours.

THINGS TO DO AND THINK ABOUT

1   Acceleration is defined as the rate of change of velocity.
An object is travelling in a straight line. The velocity, v m/s, of this object, t seconds after the start of the motion, is given by $v(t) = 8\cos\left(2t - \frac{\pi}{2}\right)$.

   (a) Find a formula for $a(t)$, the acceleration of this object, t seconds after the start of the motion.   3

   (b) Determine whether the velocity of the object is increasing or decreasing when $t = 10$.   2

3 GEOMETRY

STRAIGHT LINES 1 APPLICATIONS

DON'T FORGET

Gradient is 'change in y divided by change in x' or 'vertical change over horizontal change'.

DON'T FORGET

Avoid approximating gradients to decimals.

DON'T FORGET

Midpoint formula
$\left(\frac{x_1 + x_2}{2}, \frac{y_1 + y_2}{2}\right)$

DON'T FORGET

Distance formula
$d = \sqrt{(x_2 - x_1)^2 + (y_2 - y_1)^2}$

DON'T FORGET

To find the **equation of a straight line** you need a **point** and a **gradient**. Alternatively, two points, which will give you the gradient to use with one of the points.

WHAT YOU SHOULD ALREADY KNOW ABOUT STRAIGHT LINES

Using the gradient formula

The gradient formula is $m = \frac{y_2 - y_1}{x_2 - x_1}$

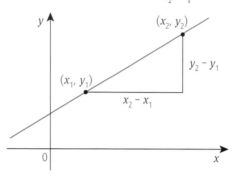

Finding the midpoint

In simple cases, the midpoint can be found by inspection.

For the line PQ, the x-coordinate $x_M = 5$, since 5 is halfway between 2 and 8, and $y_M = 1$, since 1 is midway between 7 and −5. So, the midpoint M is the point (5, 1).

Alternatively, using the midpoint formula,

$(x_M, y_M) = \left(\frac{2 + 8}{2}, \frac{7 + (-5)}{2}\right) = (5, 1)$

Calculating the distance between two points

In many cases this can be done once again by inspection, especially with a good sketch. We are just applying Pythagoras' theorem.

Alternatively, using the distance formula,

$|PQ| = \sqrt{(8 - 2)^2 + (-5 - 7)^2} = \sqrt{180} = \sqrt{36}\sqrt{5} = 6\sqrt{5}$

Using the formula $y - b = m(x - a)$

For PQ, the gradient is $m_{PQ} = \frac{7 - (-5)}{2 - 8} = -2$.

Using point P(2, 7), the equation for PQ is $y - 7 = -2(x - 2)$.

This simplifies to $y = -2x + 11$
or $y + 2x = 11$ or $y + 2x - 11 = 0$.

Quite often, you will need to use a simplified form within a longer question, such as finding where two lines meet.

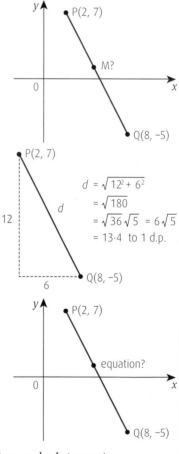

$y = mx + c$ is the equation of a straight line with **gradient m**, and **y-intercept c**.

$y - b = m(x - a)$ is the formula for finding the equation of a straight line with **gradient m**, passing through the **point (a, b)**.

Finding the point of intersection of two lines

Solve the equations of the two lines simultaneously:

$y + 2x = 11$

$3y - 2x = 1$

$4y = 12 \Rightarrow y = 3$

$3 + 2x = 11 \Rightarrow 2x = 8 \Rightarrow x = 4$

$x = 4, y = 3 \Rightarrow$ point of intersection is $(4, 3)$

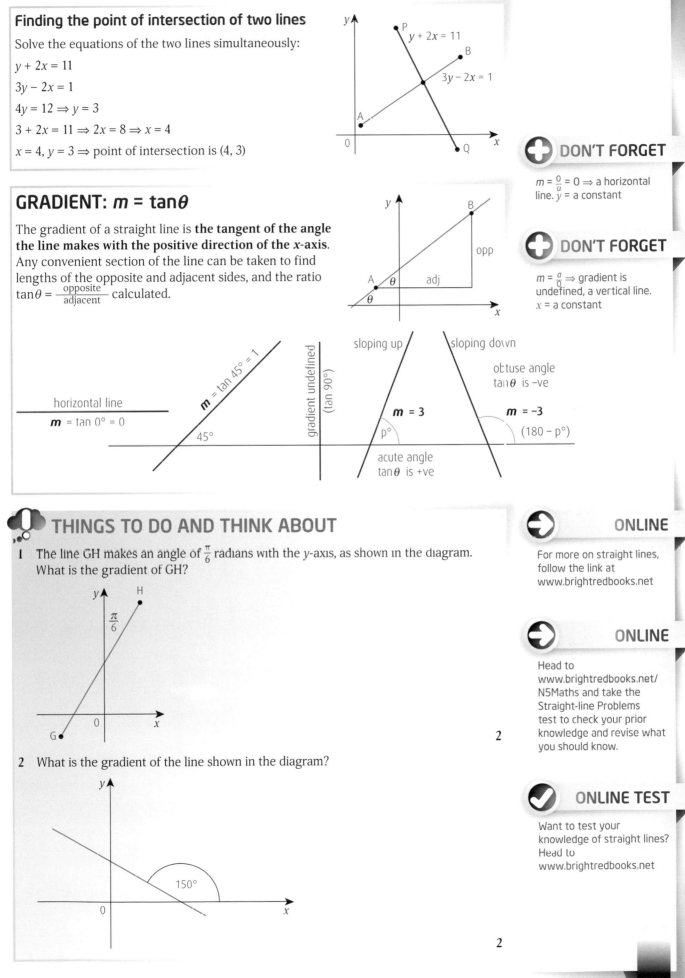

GRADIENT: $m = \tan\theta$

The gradient of a straight line is **the tangent of the angle the line makes with the positive direction of the x-axis**. Any convenient section of the line can be taken to find lengths of the opposite and adjacent sides, and the ratio $\tan\theta = \frac{\text{opposite}}{\text{adjacent}}$ calculated.

DON'T FORGET

$m = \frac{0}{a} = 0 \Rightarrow$ a horizontal line. $y =$ a constant

DON'T FORGET

$m = \frac{a}{0} \Rightarrow$ gradient is undefined, a vertical line. $x =$ a constant

THINGS TO DO AND THINK ABOUT

1 The line GH makes an angle of $\frac{\pi}{6}$ radians with the y-axis, as shown in the diagram. What is the gradient of GH?

2 What is the gradient of the line shown in the diagram?

ONLINE

For more on straight lines, follow the link at www.brightredbooks.net

ONLINE

Head to www.brightredbooks.net/N5Maths and take the Straight-line Problems test to check your prior knowledge and revise what you should know.

ONLINE TEST

Want to test your knowledge of straight lines? Head to www.brightredbooks.net

2

2

STRAIGHT LINES 2  APPLICATIONS

PARALLEL LINES AND COLLINEARITY

Parallel lines have the same gradient.

gradient = m_1

gradient = m_2

$m_1 = m_2 \Rightarrow$ lines are parallel.

Collinear points all lie on the same straight line.

Example: 1

Show that the points P(–2, –6), Q(0, –5) and R(6, –2) are collinear.

DON'T FORGET

In order to show collinearity, statements should include mention of a common point.

Solution:

$m_{PQ} = \frac{-5 - (-6)}{0 - (-2)} = \frac{1}{2}$, $m_{QR} = \frac{-2 - (-5)}{6 - 0} = \frac{1}{2}$

$m_{PQ} = m_{QR} \Rightarrow$ PQ and QR are parallel and since Q is a common point, points P, Q and R are collinear.

DON'T FORGET

To find a perpendicular gradient you change the sign and invert.

PERPENDICULAR LINES

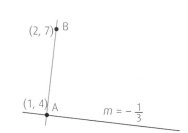

$m_1 = \frac{4}{5}$

$m_2 = ?$

(2, 7) B

(1, 4) A

$m = -\frac{1}{3}$

DON'T FORGET

Clearly label gradients to identify which line they apply to, for instance m_{AB}

If two lines are perpendicular then $m_1 m_2 = -1$.

Here, m_2 will be the negative reciprocal of m_1. Invert the fraction $\frac{4}{5}$ and change its sign to get $m_2 = -\frac{5}{4}$.

In the second case, in order to show that the two lines are perpendicular, work out the unknown gradient: $m_{AB} = \frac{7 - 4}{2 - 1} = 3$.

Then use the property $m_1 m_2 = -1$ and state

'Since $3 \times -\frac{1}{3} = -1$, the lines are perpendicular.'

If the labelling m_1 and m_2 is not used, your answer should not just say '$m_1 m_2 = -1$, hence lines are perpendicular'.

PERPENDICULAR BISECTOR

A line which cuts another line in half at right angles is a **perpendicular bisector**.

Example: 2

Find the equation of the perpendicular bisector of PQ.

contd

Solution:

The perpendicular bisector of a line PQ is
- perpendicular to PQ
- passes through the midpoint of PQ.

We already have $m_{PQ} = -2$ (see p 58) and the midpoint, M, is (5, 1).

The gradient of the perpendicular will be $\frac{1}{2}$, since $-2 \times \frac{1}{2} = -1$

Using $y - b = m(x - a)$, the equation of the perpendicular bisector of PQ is

$$y - 1 = \tfrac{1}{2}(x - 5)$$
$$2y - 2 = x - 5$$
$$2y = x - 3$$

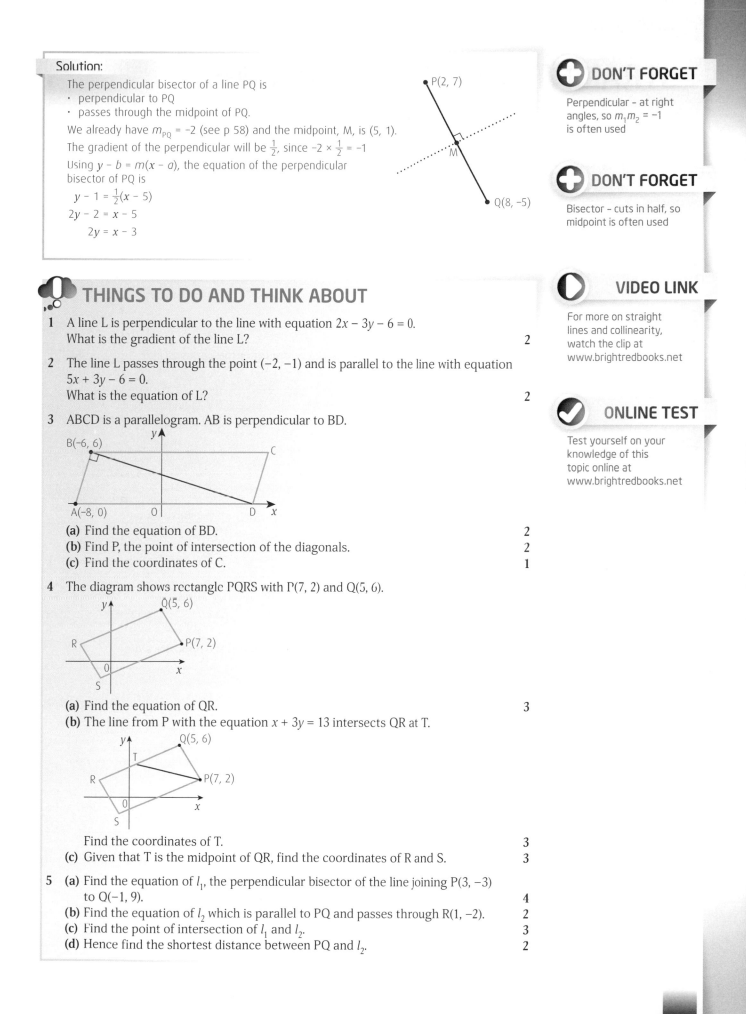

DON'T FORGET

Perpendicular – at right angles, so $m_1 m_2 = -1$ is often used

DON'T FORGET

Bisector – cuts in half, so midpoint is often used

THINGS TO DO AND THINK ABOUT

1   A line L is perpendicular to the line with equation $2x - 3y - 6 = 0$.
What is the gradient of the line L?   2

2   The line L passes through the point (−2, −1) and is parallel to the line with equation $5x + 3y - 6 = 0$.
What is the equation of L?   2

3   ABCD is a parallelogram. AB is perpendicular to BD.

(a) Find the equation of BD.   2
(b) Find P, the point of intersection of the diagonals.   2
(c) Find the coordinates of C.   1

4   The diagram shows rectangle PQRS with P(7, 2) and Q(5, 6).

(a) Find the equation of QR.   3
(b) The line from P with the equation $x + 3y = 13$ intersects QR at T.

Find the coordinates of T.   3
(c) Given that T is the midpoint of QR, find the coordinates of R and S.   3

5   (a) Find the equation of l_1, the perpendicular bisector of the line joining P(3, −3) to Q(−1, 9).   4
(b) Find the equation of l_2 which is parallel to PQ and passes through R(1, −2).   2
(c) Find the point of intersection of l_1 and l_2.   3
(d) Hence find the shortest distance between PQ and l_2.   2

VIDEO LINK

For more on straight lines and collinearity, watch the clip at www.brightredbooks.net

ONLINE TEST

Test yourself on your knowledge of this topic online at www.brightredbooks.net

STRAIGHT LINES 3 ◖APPLICATIONS◗

DON'T FORGET ➕

You need your knowledge of the basic properties of triangles and quadrilaterals.

DON'T FORGET ➕

In geometric problems, a good sketch can help you decide what to do.

DON'T FORGET ➕

Straight lines will intersect each other unless they are parallel.

LINES IN TRIANGLES

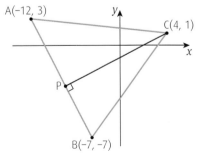

Medians

A **median** is a line joining a **vertex** to the **midpoint** of the **opposite side**.

Example: 1

Triangle ABC has vertices A(−12, 3), B(−7, −7) and C(4, 1). Find the equation of the median BM.

A(−12, 3)   M   C(4, 1)

B(−7, −7)

Solution:

For BM, find M, the midpoint of AC:

$M\left(\frac{-12 + 4}{2}, \frac{3 + 1}{2}\right) = M(-4, 2)$.

Now find the gradient of

BM: $m_{BM} = \frac{-7 - 2}{-7 + 4} = 3$.

So, equation of BM is $y - (-7) = 3(x - (-7))$ (using point B, although point M could be used) which gives

$y = 3x + 14$.

Altitudes

An **altitude** is a line through a **vertex perpendicular** to the **opposite side**.

Example: 2

Triangle ABC has vertices A(−12, 3), B(−7, −7) and C(4, 1). Find the equation of the altitude CP.

A(−12, 3)   C(4, 1)

P

B(−7, −7)

Solution:

For the gradient of CP, first find the gradient of AB: $m_{AB} = \frac{3 + 7}{-12 + 7} = -2$

So, the gradient of CP, $m_{CP} = \frac{1}{2}$, (using $m_{AB} \times m_{CP} = -1$).

Therefore, equation of CP is $y - 1 = \frac{1}{2}(x - 4)$, (using point C), giving $2y = x - 2$.

Solving the equations of the two lines, in Examples 1 and 2, simultaneously gives (−6, −4). This is the intersection of the median BM and the altitude CP.

Concurrency

Each set of medians, altitudes, perpendicular bisectors and angle bisectors for any triangle are **concurrent**.

contd

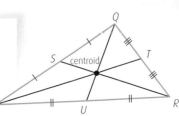

The three **medians** pass through a point called the **centroid**. The centroid is exactly two-thirds of the way along each median.

The three **altitudes** pass through a point called the **orthocentre**. The orthocentre can be inside (for an acute triangle) or outside (for an obtuse triangle).

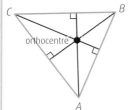

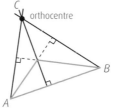

Orthocentre is inside the triangle.   Orthocentre is outside the triangle.

The three **perpendicular bisectors** meet at a point called the **circumcentre**. This will be outside the triangle if the triangle is obtuse.

A line that cuts an angle in half is the **angle bisector**. The three angle bisectors meet at a point called the **incentre**.

It's not very likely that three random lines will be concurrent. If they are parallel, they won't even meet. Even when none are parallel, it's more likely they will look like diagram 1 than diagram 2 here:

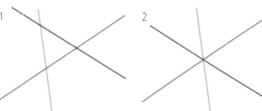

ONLINE

To see examples of a circumcentre and incentre, head to www.brightredbooks.net

Example: 3

Are these three lines concurrent?

$2x + y - 2 = 0$ (1)
$x - y + 5 = 0$ (2)
$x - 3y + 13 = 0$ (3)

Solution:

To solve this, first find the point of intersection of two of the lines. We can see from the coefficients that none of the lines are parallel, so any two must intersect if you extend them far enough. Then, see whether that point also lies on the third line.

Adding equations 1 and 2 gives: $3x + 3 = 0 \Rightarrow x = -1$

Substitution in equation 1 gives: $-2 + y - 2 = 0 \Rightarrow y = 4$ so the point of intersection is $(-1, 4)$

Now, test to see whether $(-1, 4)$ satisfies equation 3:
$x - 3y + 13 = -1 - 12 + 13 = 0$

Yes it does, so **the three lines are concurrent**, and the point of concurrency is $(-1, 4)$.

ONLINE

The centroid, orthocentre and circumcentre for a triangle are collinear and lie on a line called the **Euler line**. More information and other interesting facts about the Euler line and triangle centres can be found at www.brightredbooks.net

THINGS TO DO AND THINK ABOUT

1   Triangle PQR has vertices at P(−3, −2), Q(−1, 4) and R(3, 6). PS is a median. What is the gradient of PS?

2   Triangle PQR has vertex P on the x-axis, as shown in the diagram.

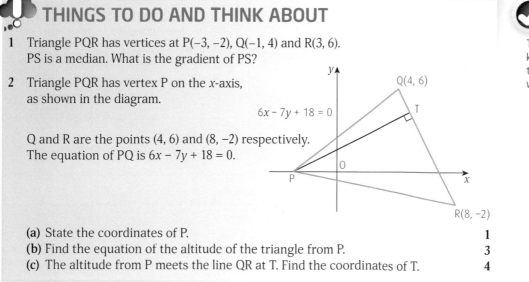

Q and R are the points (4, 6) and (8, −2) respectively. The equation of PQ is $6x − 7y + 18 = 0$.

(a) State the coordinates of P.   1
(b) Find the equation of the altitude of the triangle from P.   3
(c) The altitude from P meets the line QR at T. Find the coordinates of T.   4

ONLINE TEST

Test yourself on your knowledge of this topic online at www.brightredbooks.net

CIRCLES 1 APPLICATIONS

EQUATION OF A CIRCLE

Although the natural place to learn about circles is within geometry, it should be noted that many algebraic skills are applied when tackling problems involving circles.

$$(x - a)^2 + (y - b)^2 = r^2$$

is the equation of a circle with centre (a, b) and radius r.

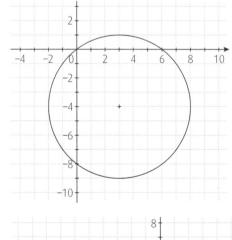

If the centre is at the origin, (a, b) is $(0, 0)$, so the equation of the circle is

$$x^2 + y^2 = r^2.$$

The **unit circle** has centre the origin and radius 1:
$$x^2 + y^2 = 1.$$

The general equation of the circle is

$$x^2 + y^2 + 2gx + 2fy + c = 0.$$

It represents a circle with centre $(-g, -f)$ and radius $\sqrt{g^2 + f^2 - c}$.

For a circle to exist, $g^2 + f^2 - c > 0$.

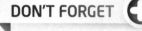

DON'T FORGET

You should know these equations and use the formula sheet if you are unsure in the examination.

DON'T FORGET

$r > 0$ for the circle to exist.

Example: 1

Sketch the circles represented by these equations

(a) $(x - 3)^2 + (y + 4)^2 = 25$

(b) $x^2 + y^2 + 10x - 4y + 25 = 0$

Solution:

(a) Centre is $(3, -4)$, radius 5.
By substituting $x = 0$, then $y = 0$, you can quickly find where the circle crosses the y and x-axes, respectively.

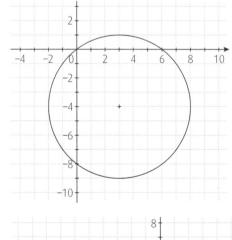

b) $2g = 10$, $2f = -4 \Rightarrow -g = -5$, $-f = 2$

Centre is $(-5, 2)$,
radius $\sqrt{5^2 + (-2)^2 - 25} = 2$.

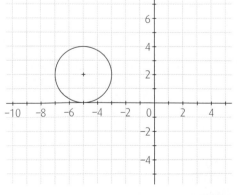

contd

Example: 2

Does the equation $x^2 + y^2 - 8x + 6y + 30 = 0$ represent a circle?

Solution:

For a circle to exist, $g^2 + f^2 - c > 0$.

$g = -4, f = 3, (-4)^2 + 3^2 - 30 = -5$.

Since $g^2 + f^2 - c < 0$, the equation $x^2 + y^2 - 8x + 6y + 30 = 0$ does not represent a circle.

Example: 3

Points P $(-6, -3)$ and Q $(2, 3)$ are at either end of the diameter of a circle.

What is the equation of this circle?

Solution:

Radius $= \frac{1}{2} \times$ diameter. Distance PQ $= \sqrt{(2 - (-6))^2 + (3 - (-3))^2} = 10$, so radius is 5.

Centre of circle is midpoint of PQ. Midpoint $= \left(\frac{-6 + 2}{2}, \frac{-3 + 3}{2}\right) = (-2, 0)$

Substituting into $(x - a)^2 + (y - b)^2 = r^2$ gives $(x - (-2))^2 + (y - 0)^2 = 5^2$

So the equation of the circle is $(x + 2)^2 + y^2 = 25$.

Example: 4

C_1 and C_2 are congruent circles.

C_1 has equation $x^2 + y^2 - 4x - 30 = 0$.

C_1 touches C_2 at T $(5, 5)$.

What is the equation of C_2?

Solution:

C_1 has centre $(2, 0)$. The radius is the distance from $(2, 0)$ to T $(5, 5) = \sqrt{34}$.

The journey from centre C_1 to T $= \binom{5}{5} - \binom{2}{0} = \binom{3}{5}$.

This is the same as the journey from T to C_2.

$\binom{5}{5} + \binom{3}{5} = \binom{8}{10} \Rightarrow$ centre C_2 is $(8, 10)$.

Circle C_2 has equation $(x - 8)^2 + (y - 10)^2 = 34$.

Here a 'vectors method' was used; alternatively use 'stepping out', but be sure to include a fully-annotated sketch.

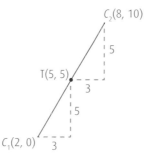

THINGS TO DO AND THINK ABOUT

1. Work out the radius and the coordinates of the centre of the circle with equation $4x^2 + 4y^2 - 16x + 8y + 11 = 0$ **2**

2. Not every equation which **looks** like it represents a circle actually does represent a circle. For example, $x^2 + y^2 + 2x - 6y + 50 = 0$ **isn't** the equation of a circle. Why not?   **2**

3. Does A$(-3, 1)$ lie on, or outside, the circle with equation $x^2 + y^2 - 4x - 10y + 4 = 0$?   **2**

4. Given that the equation $x^2 + y^2 - 2px - 4py + 3p + 2 = 0$ represents a circle, determine the range of values of p.   **5**

CIRCLES 2 — APPLICATIONS

INTERSECTING CIRCLES

To determine whether or not two circles **intersect** at two points, one point or not at all
- **add** the lengths of the **two radii** together $(r_1 + r_2)$
- **compare** with the **distance between the centres** (d).

The circles meet **externally** at **one point**, $d = r_1 + r_2$.

The circles **intersect** at **two points** (they overlap), $d < r_1 + r_2$.

The circles meet **internally** at **one point**, $d = r_1 - r_2$.

The circles **do not touch**, $d > r_1 + r_2$.

The circles **do not touch** and one circle is contained **within** the other, $d < r_1 - r_2$.

The circles **do not touch**, one circle is contained **within** the other and they are **concentric**, $d < r_1 - r_2$.

DON'T FORGET

Concentric circles have the same centre.

DON'T FORGET

The line through the centres is a line of symmetry for each pair of circles.

DON'T FORGET

You must clearly communicate the condition you are using and any resulting conclusion.

Example: 1

Do the circles with equations $(x + 2)^2 + (y - 5)^2 = 16$ and $x^2 + y^2 - 6x + 14y - 42 = 0$ intersect?

Solution:

$(x + 2)^2 + (y - 5)^2 = 16$ has centre $(-2, 5)$ and radius $r_1 = 4$.

$x^2 + y^2 - 6x + 14y - 42 = 0$ has centre $(3, -7)$ and radius $r_2 = \sqrt{(-3)^2 + 7^2 - (-42)} = 10$.

The distance between the centres $d = \sqrt{(3 - (-2))^2 + ((-7) - 5)^2} = 13$

$r_1 + r_2 = 14$.

Since $d < r_1 + r_2$, the circles intersect at two points.

LINES AND CIRCLES

A line may cut a circle twice, or just touch the circle at one point or not meet the circle at all.

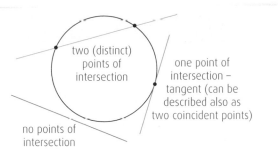

two (distinct) points of intersection

one point of intersection – tangent (can be described also as two coincident points)

no points of intersection

Example: 2

Find any points of intersection of the line $x = 6$ and the circle $x^2 + y^2 - 4x + 10y + 4 = 0$.

Solution:

Substitute $x = 6$ into the equation for the circle to get

$36 + y^2 - 24 + 10y + 4 = 0$

$y^2 + 10y + 16 = 0 \Rightarrow (y + 2)(y + 8) = 0 \Rightarrow y = -2, y = -8$

giving the two points where the line cuts the circle, $(6, -2)$ and $(6, -8)$.

Tangency

When a line just touches a curve, that is **one point of intersection** (or two coincident points), then that line is a **tangent**.

DON'T FORGET

$b^2 - 4ac = 0 \Rightarrow$ equal roots.

Example: 3

Show that the line $y = -2x - 3$ is a tangent to the circle with equation $x^2 + y^2 + 2x + 12y + 32 = 0$.

Solution:

Substitute $y = -2x - 3$ into the equation for the circle to get

$x^2 + (-2x - 3)^2 + 2x + 12(-2x - 3) + 32 = 0$

$x^2 + 4x^2 + 12x + 9 + 2x - 24x - 36 + 32 = 0 \Rightarrow 5x^2 - 10x + 5 = 0 \Rightarrow 5(x^2 - 2x + 1) = 0$

Using the discriminant we get:

$b^2 - 4ac = (-2)^2 - 4 \times 1 \times 1 = 0 \Rightarrow$ equal roots, so the line is a tangent.

An alternative method for Example 3 is to factorise to find two real, equal roots.

ONLINE

Head to www.brightredbooks.net to see an extra worked example and to complete further questions on this topic.

THINGS TO DO AND THINK ABOUT

1   Relative to a suitable set of coordinate axes, diagram 1 shows the line $2x - y + 5 = 0$ intersecting the circle $x^2 + y^2 - 6x - 2y - 30 = 0$ at the points P and Q.

(a) Find the coordinates of P and Q.   6

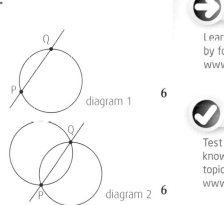

diagram 1

Diagram 2 shows the circle from part (a) and a second congruent circle, which also passes through P and Q.

(b) Determine the equation of this second circle.   6

diagram 2   6

ONLINE

Learn more about this by following the link at www.brightredbooks.net

ONLINE TEST

Test yourself on your knowledge of this topic online at www.brightredbooks.net

VECTORS 1 — EXPRESSIONS AND FUNCTIONS

BASIC PROPERTIES OF VECTORS

A **vector** is a quantity with both **magnitude** and **direction**. It can be represented by a **directed line segment**.

In the diagram below, E (5, 3, 4) is a point in three-dimensional space.

$\mathbf{e} = \begin{pmatrix} 5 \\ 3 \\ 4 \end{pmatrix}$ is a vector and must be written with the components vertical.

In the diagram below, the position vector **e** is the directed line segment $\overrightarrow{OE}$, and gives the components of the move from O, the origin, to E. **e** is the **position vector** of E because it fixes its position. There can only be one point E. However, there can be any number of representations of the vector **e** because every vector with the same components is an identical move or journey, but starting in a different place in space.

ONLINE TEST

Test yourself on your knowledge of this topic online at www.brightredbooks.net

DON'T FORGET

Use the rule $\overrightarrow{AB} = \mathbf{b} - \mathbf{a}$ to find the components of a vector given the coordinates of A and B.

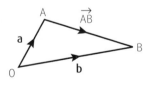

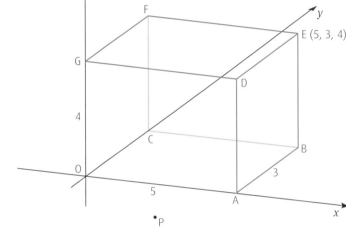

This diagram of a cuboid is referred to in several examples in this chapter.

The cuboid pictured has been drawn on a coordinate diagram with x, y and z-axes in three dimensions at right angles to each other. Point P lies under the cuboid.

E is the point (5, 3, 4), so OA = 5 units, OC = 3 units and OG = 4 units in length.

An important and useful result: $\overrightarrow{AB} = \mathbf{b} - \mathbf{a}$.

Example: 1

In the diagram of the cuboid above, $\overrightarrow{PF} = \begin{pmatrix} -3 \\ 1 \\ 10 \end{pmatrix}$. Find the coordinates of P.

DON'T FORGET

Coordinates of points (a, b, c). Vectors $\begin{pmatrix} a \\ b \\ c \end{pmatrix}$.

Solution:

F (0, 3, 4) is easily deduced from the diagram.

Since $\overrightarrow{PF} = \mathbf{f} - \mathbf{p}$, substitution gives us $\begin{pmatrix} -3 \\ 1 \\ 10 \end{pmatrix} = \begin{pmatrix} 0 \\ 3 \\ 4 \end{pmatrix} - \mathbf{p}$

This vector equation is easily rearranged and solved to give

$\mathbf{p} = \begin{pmatrix} 3 \\ 2 \\ -6 \end{pmatrix}$ so P is the point (3, 2, −6).

The **magnitude** (length) of a vector $\mathbf{v} = \begin{pmatrix} a \\ b \\ c \end{pmatrix}$ is given by $|\mathbf{v}| = \sqrt{a^2 + b^2 + c^2}$

Example: 2

Find $|\overrightarrow{PF}|$, the magnitude or length of $\overrightarrow{PF}$.

contd

Solution:

The magnitude is calculated using Pythagoras' theorem, extended to three dimensions.

$$\left|\overrightarrow{PF}\right| = \sqrt{(-3)^2 + 1^2 + 10^2} = \sqrt{110}$$

In general, the distance between two points (x_1, y_1, z_1) and (x_2, y_2, z_2) is

$$d = \sqrt{(x_2 - x_1)^2 + (y_2 - y_1)^2 + (z_2 - z_1)^2}$$

which you will recognise as the distance formula, extended to three dimensions.

Equal vectors have **equal lengths** and are **parallel**.

The **negative** of a vector has the **same magnitude**, but **opposite direction**.

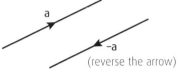

(reverse the arrow)

The **zero vector**, $0 = \begin{pmatrix} 0 \\ 0 \\ 0 \end{pmatrix}$, has **zero magnitude** and **no defined direction**.

Given vectors **u** and **v**, **u** + **v** gives the **resultant** of the vector journey **u** followed by **v**.

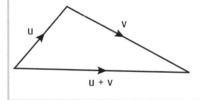

🗯 THINGS TO DO AND THINK ABOUT

1   The diagram shows a regular hexagon PQRSTW.

$\overrightarrow{PW}$ and $\overrightarrow{PQ}$ represent vectors **u** and **v** respectively.

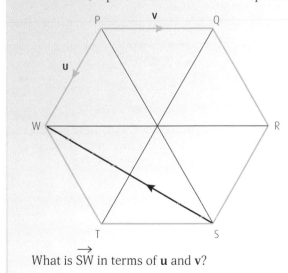

What is $\overrightarrow{SW}$ in terms of **u** and **v**?

VECTORS 2 ⬭ EXPRESSIONS AND FUNCTIONS

MIDPOINT OF A LINE, UNIT VECTORS, COLLINEARITY AND SECTION FORMULA

Midpoint of a line

Vector paths can be used to find the midpoint of a line. If M is the midpoint of AB, then the diagram shows

$\mathbf{m} = \frac{1}{2}(\mathbf{a} + \mathbf{b})$

$\mathbf{m} = \overrightarrow{OM} = \overrightarrow{OA} + \frac{1}{2}\overrightarrow{AB}$ start at O, move to A, then halfway along $\overrightarrow{AB}$.

$= \mathbf{a} + \frac{1}{2}(\mathbf{b} - \mathbf{a})$

$= \frac{1}{2}\mathbf{a} + \frac{1}{2}\mathbf{b}$

$= \frac{1}{2}(\mathbf{a} + \mathbf{b})$

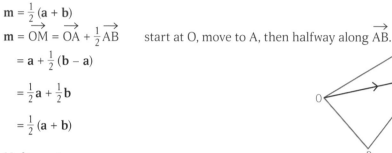

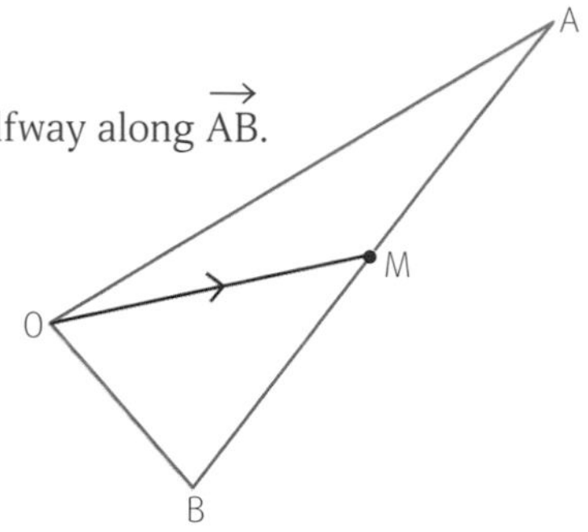

Unit vectors

Unit vectors have length 1. Examples are **i**, **j** and **k**, which take the directions of mutually perpendicular axes, x, y and z.

$\mathbf{i} = \begin{pmatrix} 1 \\ 0 \\ 0 \end{pmatrix}$ $\mathbf{j} = \begin{pmatrix} 0 \\ 1 \\ 0 \end{pmatrix}$ $\mathbf{k} = \begin{pmatrix} 0 \\ 0 \\ 1 \end{pmatrix}$ so that $\overrightarrow{OE} = \mathbf{e} = \begin{pmatrix} 5 \\ 3 \\ 4 \end{pmatrix} = 5\mathbf{i} + 3\mathbf{j} + 4\mathbf{k}$.

DON'T FORGET ➕

Remember $\frac{1}{2}\mathbf{a}$ is **a** with all the components halved.

Example: 1

Find a unit vector, **u**, parallel to $\overrightarrow{PF}$ in the diagram on p 68.

Solution:

u will be a scalar multiple of $\overrightarrow{PF}$ (in order to be parallel).
$\overrightarrow{PF}$ has length $\sqrt{110}$ and **u** has length 1. Dividing each component of $\overrightarrow{PF}$ by $\sqrt{110}$ will make its length 1.

$\mathbf{u} = \frac{1}{\sqrt{110}}\begin{pmatrix} -3 \\ 1 \\ 10 \end{pmatrix}$ (no need to evaluate further unless the question requires it).

Collinearity

If you found that two lines had the **same gradient**, you would know they were **parallel**. If you also knew that a particular point lay on both lines, then the two lines would no longer be separate lines but would both be parts of the same straight line.

DON'T FORGET ➕

parallel lines + common point ⟹ collinearity

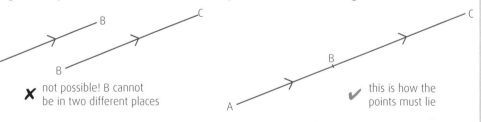

✗ not possible! B cannot be in two different places

✔ this is how the points must lie

DON'T FORGET ➕

'parallel vectors' **and** 'common point' must be mentioned or marks will be lost.

The vector work in this section does the same thing, but can be used in two and three dimensions. To show that lines are parallel using vectors, show that one is a **scalar multiple** of the other. If the lines/vectors also have a point in common, then they are one line and all the points lie on the same line, so they are collinear.

Example: 2

Show that A(3, –1, 5), B(6, 0, 3) and C(15, 3, –3) are collinear points, and draw a diagram to show their relative positions.

contd

Solution:

Choose any two directed line segments using these points.

Choosing $\vec{AB}$ and $\vec{AC}$:

$$\vec{AB} = \begin{pmatrix} 6-3 \\ 0+1 \\ 3-5 \end{pmatrix} = \begin{pmatrix} 3 \\ 1 \\ -2 \end{pmatrix} \text{ and } \vec{AC} = \begin{pmatrix} 15-3 \\ 3+1 \\ -3-5 \end{pmatrix} = \begin{pmatrix} 12 \\ 4 \\ -8 \end{pmatrix} = 4\begin{pmatrix} 3 \\ 1 \\ -2 \end{pmatrix}$$

$\vec{AC} = 4\vec{AB}$ so $\vec{AC}$ is a scalar multiple of $\vec{AB}$. Hence AC is parallel to AB and, since A is a common point, the points A, B and C are collinear.

Next, draw a line and mark the points A, B and C on it so that AC is 4 times the length of AB.

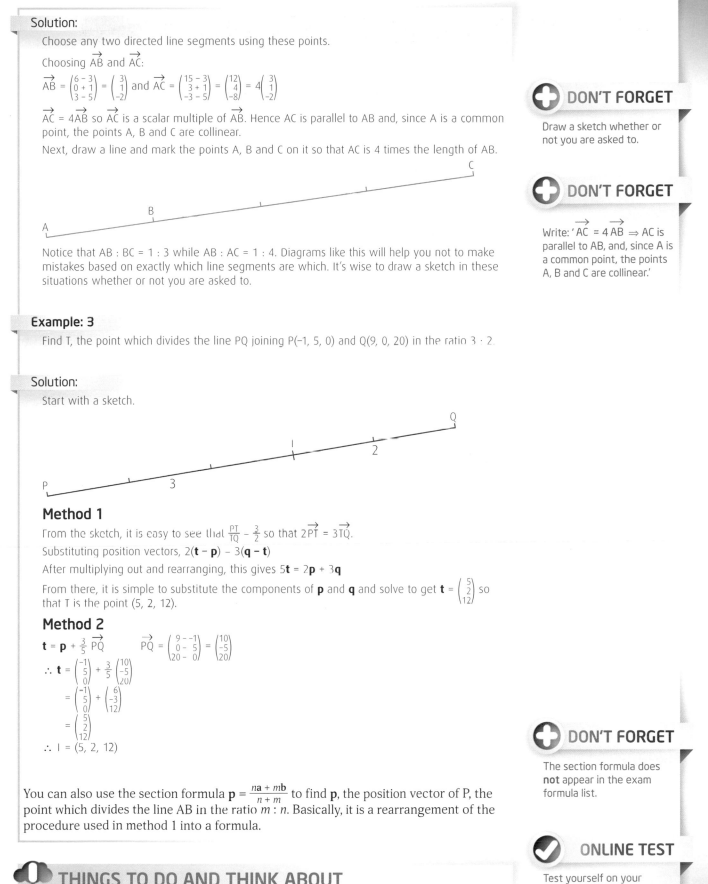

Notice that AB : BC = 1 : 3 while AB : AC = 1 : 4. Diagrams like this will help you not to make mistakes based on exactly which line segments are which. It's wise to draw a sketch in these situations whether or not you are asked to.

Example: 3

Find T, the point which divides the line PQ joining P(–1, 5, 0) and Q(9, 0, 20) in the ratio 3 : 2.

Solution:

Start with a sketch.

Method 1

From the sketch, it is easy to see that $\frac{PT}{TQ} = \frac{3}{2}$ so that $2\vec{PT} = 3\vec{TQ}$.

Substituting position vectors, $2(\mathbf{t} - \mathbf{p}) = 3(\mathbf{q} - \mathbf{t})$

After multiplying out and rearranging, this gives $5\mathbf{t} = 2\mathbf{p} + 3\mathbf{q}$

From there, it is simple to substitute the components of $\mathbf{p}$ and $\mathbf{q}$ and solve to get $\mathbf{t} = \begin{pmatrix} 5 \\ 2 \\ 12 \end{pmatrix}$ so that T is the point (5, 2, 12).

Method 2

$\mathbf{t} = \mathbf{p} + \frac{3}{5}\vec{PQ}$ $\vec{PQ} = \begin{pmatrix} 9--1 \\ 0-5 \\ 20-0 \end{pmatrix} = \begin{pmatrix} 10 \\ -5 \\ 20 \end{pmatrix}$

$\therefore \mathbf{t} = \begin{pmatrix} -1 \\ 5 \\ 0 \end{pmatrix} + \frac{3}{5}\begin{pmatrix} 10 \\ -5 \\ 20 \end{pmatrix}$

$= \begin{pmatrix} -1 \\ 5 \\ 0 \end{pmatrix} + \begin{pmatrix} 6 \\ -3 \\ 12 \end{pmatrix}$

$= \begin{pmatrix} 5 \\ 2 \\ 12 \end{pmatrix}$

$\therefore$ T = (5, 2, 12)

You can also use the section formula $\mathbf{p} = \dfrac{n\mathbf{a} + m\mathbf{b}}{n + m}$ to find $\mathbf{p}$, the position vector of P, the point which divides the line AB in the ratio $m : n$. Basically, it is a rearrangement of the procedure used in method 1 into a formula.

THINGS TO DO AND THINK ABOUT

1 If, in Example 3 above, you were asked to find S where PS = 2PQ, you would find a point lying outside the line PQ. Try sketching it and working out the coordinates of S.

DON'T FORGET

Draw a sketch whether or not you are asked to.

DON'T FORGET

Write: '$\vec{AC}$ = 4 $\vec{AB}$ $\Rightarrow$ AC is parallel to AB, and, since A is a common point, the points A, B and C are collinear.'

DON'T FORGET

The section formula does **not** appear in the exam formula list.

ONLINE TEST

Test yourself on your knowledge of this topic online at www.brightredbooks.net

VECTOR ALGEBRA AND THE SCALAR PRODUCT

The **scalar product** (**dot product**) of two vectors involves the **multiplication** of the **magnitudes** of the two vectors and the **cosine of the angle between them**.

$\mathbf{a}.\mathbf{b} = |\mathbf{a}|\,|\mathbf{b}|\cos\theta$ where θ is the angle between the vectors when they are placed nose-to-nose or tail-to-tail (both pointing in, or both pointing out).

The answer works out as **scalar**, or a number – not another vector.

Example: 1

Calculate **p.q** where $|\mathbf{p}| = 4$, $|\mathbf{q}| = 5$.

Solution:

First, rearrange the diagram so that both vectors point out from the vertex.

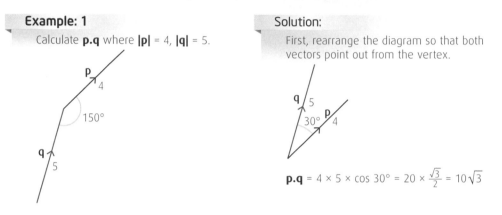

$\mathbf{p}.\mathbf{q} = 4 \times 5 \times \cos 30° = 20 \times \frac{\sqrt{3}}{2} = 10\sqrt{3}$

Particularly interesting cases are when $\theta = 0°$ or $90°$.
- $\theta = 0°$ as, for example, $\mathbf{a}.\mathbf{a} = |\mathbf{a}|\,|\mathbf{a}|\cos 0°$. Since $\cos 0° = 1$, the result is $\mathbf{a}.\mathbf{a} = |\mathbf{a}|^2$.
- $\theta = 90°$ where the vectors are perpendicular. Since $\cos 90° = 0$, the scalar product is then 0.

The second result is particularly handy in reverse. That is, if you find that a scalar product works out to be 0, the angle must be $90°$, showing that the vectors are **perpendicular**.

The scalar product can also be worked out from components.

$\mathbf{a}.\mathbf{b} = a_1b_1 + a_2b_2 + a_3b_3$ (where a_1 etc. are vector components)

Here is an example of the calculation of the scalar product of two vectors:

$\begin{pmatrix} 3 \\ -1 \\ 2 \end{pmatrix}.\begin{pmatrix} 7 \\ 4 \\ -5 \end{pmatrix} = 3 \times 7 + (-1) \times 4 + 2 \times (-5) = 21 - 4 - 10 = 7$

Example: 2

For what value of t is vector $\mathbf{p} = 3\mathbf{i} - \mathbf{j} + t\mathbf{k}$ perpendicular to vector $\mathbf{q} = 2\mathbf{i} + 3\mathbf{j} - \mathbf{k}$?

Solution:

$\mathbf{a}.\mathbf{b} = 0$ for perpendicular vectors.

$\mathbf{p}.\mathbf{q} = 3 \times 2 + -1 \times 3 + t \times -1 = 6 - 3 - t = 0 \Rightarrow t = 3$.

What is really useful is not so much these results for the scalar product separately but what we can do by combining them.

$\cos\theta = \dfrac{a_1b_1 + a_2b_2 + a_3b_3}{|\mathbf{a}|\,|\mathbf{b}|}$

Example: 3

Find angle EPF in the cuboid diagram on page 68.

contd

DON'T FORGET

Be careful to note the directions of arrows on the diagrams.

DON'T FORGET

If $|\mathbf{a}|, |\mathbf{b}| \neq 0$ then $\mathbf{a}.\mathbf{b} = 0$ if and only if the directions of **a** and **b** are at right angles.

Solution:

$$\overrightarrow{PE} = \begin{pmatrix} 2 \\ 1 \\ 10 \end{pmatrix} \quad \overrightarrow{PF} = \begin{pmatrix} -3 \\ 1 \\ 10 \end{pmatrix}$$

$\overrightarrow{PE}.\overrightarrow{PF} = 2 \times -3 + 1 \times 1 + 10 \times 10 = -6 + 1 + 100 = 95$

We already have $|PF| = \sqrt{110}$ and $|PE| = \sqrt{4 + 1 + 100} = \sqrt{105}$

$\cos E\hat{P}F = \dfrac{\overrightarrow{PE}.\overrightarrow{PF}}{|PE||PF|} = \dfrac{95}{\sqrt{105} \times \sqrt{110}} = 0.8839 \ldots$

$E\hat{P}F = \cos^{-1} 0.8839 = 27.9°$

Some questions on vectors in the examination don't involve components or coordinates. They use the scalar product rule and two other vector results which you need to remember:

a.b = b.a $\qquad$ **a.(b + c) = a.b + a.c**

These questions will normally have some information about the lengths of vectors and angles between the vectors, possibly given with the help of a diagram. Vector paths can crop up too.

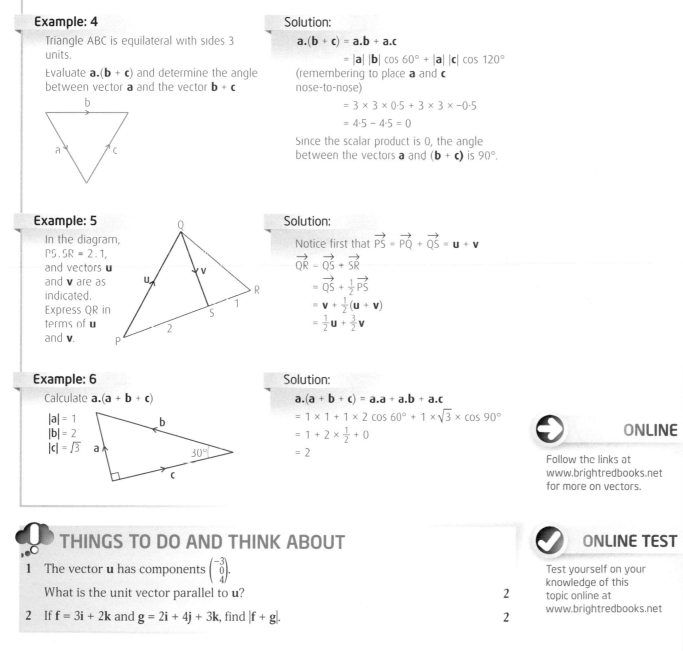

Example: 4

Triangle ABC is equilateral with sides 3 units.

Evaluate **a.(b + c)** and determine the angle between vector **a** and the vector **b + c**

Solution:

a.(b + c) = a.b + a.c

$= |\mathbf{a}| \, |\mathbf{b}| \cos 60° + |\mathbf{a}| \, |\mathbf{c}| \cos 120°$

(remembering to place **a** and **c** nose-to-nose)

$= 3 \times 3 \times 0.5 + 3 \times 3 \times -0.5$

$= 4.5 - 4.5 = 0$

Since the scalar product is 0, the angle between the vectors **a** and (**b + c**) is 90°.

Example: 5

In the diagram, PS.SR = 2.1, and vectors **u** and **v** are as indicated. Express QR in terms of **u** and **v**.

Solution:

Notice first that $\overrightarrow{PS} = \overrightarrow{PQ} + \overrightarrow{QS} = \mathbf{u} + \mathbf{v}$

$\overrightarrow{QR} = \overrightarrow{QS} + \overrightarrow{SR}$

$\qquad = \overrightarrow{QS} + \tfrac{1}{2}\overrightarrow{PS}$

$\qquad = \mathbf{v} + \tfrac{1}{2}(\mathbf{u} + \mathbf{v})$

$\qquad = \tfrac{1}{2}\mathbf{u} + \tfrac{3}{2}\mathbf{v}$

Example: 6

Calculate **a.(a + b + c)**

$|\mathbf{a}| = 1$
$|\mathbf{b}| = 2$
$|\mathbf{c}| = \sqrt{3}$

Solution:

a.(a + b + c) = a.a + a.b + a.c

$= 1 \times 1 + 1 \times 2 \cos 60° + 1 \times \sqrt{3} \times \cos 90°$

$= 1 + 2 \times \tfrac{1}{2} + 0$

$= 2$

ONLINE

Follow the links at www.brightredbooks.net for more on vectors.

THINGS TO DO AND THINK ABOUT

1 The vector **u** has components $\begin{pmatrix} -3 \\ 0 \\ 4 \end{pmatrix}$.

   What is the unit vector parallel to **u**? $\qquad$ 2

2 If $\mathbf{f} = 3\mathbf{i} + 2\mathbf{k}$ and $\mathbf{g} = 2\mathbf{i} + 4\mathbf{j} + 3\mathbf{k}$, find $|\mathbf{f} + \mathbf{g}|$. $\qquad$ 2

ONLINE TEST

Test yourself on your knowledge of this topic online at www.brightredbooks.net

4 TRIGONOMETRY

BASICS: RADIANS AND TRIGONOMETRIC RATIOS
EXPRESSIONS AND FUNCTIONS

DON'T FORGET

Think of these two ways of measuring angles as similar to using km or miles for measuring lengths. In any one case you must stick to one or the other – do not mix them up – but occasionally it is helpful to be able to switch between them.

DON'T FORGET

Make sure you are aware of whether a question requires degrees or radians. The lack of a degree sign, °, usually indicates radians, as does a range involving π, such as $0 \le x \le 2\pi$.

DON'T FORGET

degrees $\xrightarrow[\times \frac{180}{\pi}]{\times \frac{\pi}{180}}$ radians

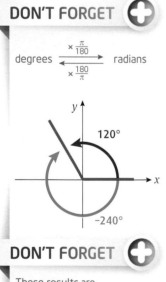

DON'T FORGET

These results are often remembered as **SOHCAHTOA**.

RADIANS

Angles are measured in either degrees or radians.

You should already know that one complete revolution = 360°.

1 radian is the angle subtended at the centre of a circle by an arc of the circle equal in length to the circle's radius. Since the circumference of a circle can be expressed as $C = 2\pi r$, it takes 2π radians to complete a full circle.

So 2π radians = 360°, or π radians = 180°.

Radians are very suitable for trigonometric calculations in the real-life uses of trigonometry which scientists and engineers work with, and are essential for work with calculus.

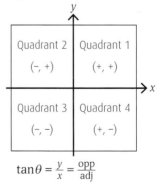

Degrees	0°	30°	45°	60°	90°	180°	270°	360°
Radians	0	$\frac{\pi}{6}$	$\frac{\pi}{4}$	$\frac{\pi}{3}$	$\frac{\pi}{2}$	π	$\frac{3\pi}{2}$	2π

TRIGONOMETRIC RATIOS

The word trigonometry comes from the Greek words 'trigon' (triangle) and 'metron' (measure), and literally means the measuring (of angles and sides) of triangles. The familiar trigonometric ratios of sine (sin), cosine (cos) and tangent (tan) come from measuring in triangles, but trigonometry is much more than that. It is used by engineers to design bridges, and by satellite navigation systems, amongst other things.

The Cartesian axes divide a plane into **four quadrants**.

Angles are usually measured from the **positive x-axis**.

Those measured **anticlockwise** are **positive**, whilst those measured **clockwise** are **negative**.

If we consider an acute angle of a right-angled triangle, we get the familiar trigonometric ratios

$$\sin \theta = \frac{y}{r} = \frac{\text{opp}}{\text{hyp}} \qquad \cos \theta = \frac{x}{r} = \frac{\text{adj}}{\text{hyp}} \qquad \tan\theta = \frac{y}{x} = \frac{\text{opp}}{\text{adj}}$$

	y	
Quadrant 2		Quadrant 1
$(-, +)$		$(+, +)$
Quadrant 3		Quadrant 4
$(-, -)$		$(+, -)$

The trigonometric ratio of any angle can then be obtained by determining
- the **quadrant** connected with the angle,
- the **sign** of x and/or y within that quadrant and
- the associated **acute angle** made with the x-axis.

r is always taken to be positive.

contd

Example: 1

Find the value of

(a) sin 120°

(b) $\cos\left(\frac{5\pi}{4}\right)$

Solution:

(a)

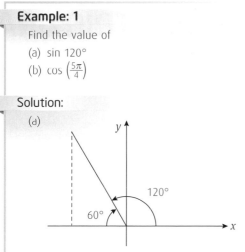

120° lies in the 2nd quadrant, in which y is positive (and x is negative).

Associated acute angle = 60° $\Rightarrow$ sin 120° = $\frac{y}{r}$ = sin 60° = $\frac{\sqrt{3}}{2}$

(b)

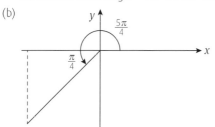

$\frac{5\pi}{4}$ lies in the 3rd quadrant, in which both x and y are negative.

Associated acute angle = $\frac{\pi}{4}$ $\Rightarrow$ $\cos\left(\frac{5\pi}{4}\right)$ = $-\frac{x}{r}$ = $-\cos\left(\frac{\pi}{4}\right)$ = $-\frac{\sqrt{2}}{2}$

A useful aid is the **CAST** diagram, which shows which trigonometric ratios are **positive** in each quadrant (and hence, by process of elimination, which must be **negative** in each quadrant too).

Sin	All
Tan	Cos

So, for example, $\cos\left(\frac{5\pi}{4}\right)$ $\Rightarrow$ 3rd quadrant $\Rightarrow$ only tan is positive

$\Rightarrow$ $\cos\left(\frac{5\pi}{4}\right)$ = $-\cos\left(\frac{\pi}{4}\right)$ = $-\frac{\sqrt{2}}{2}$.

DON'T FORGET

Mnemonics such as
'**A S**mart **T**rig **C**lass' and
'**A**ll **S**tudents **T**ake **C**alculus'
are helpful to remember the
quadrants in order, though
'CAST' can also be used for
obvious reasons.

ONLINE

For more on radians, head
to www.brightredbooks.net

THINGS TO DO AND THINK ABOUT

1   If $0 < a < 90$, which of the following is equivalent to $\cos(270 - a)°$?

A $\cos a°$

B $\sin a°$

C $-\cos a°$

D $-\sin a°$

ONLINE TEST

Head to
www.brightredbooks.net to
test yourself on the basics.

2

BASICS: EXACT VALUES AND BASIC TRIG GRAPHS

EXPRESSIONS AND FUNCTIONS

EXACT VALUES

DON'T FORGET

Rationalising the denominator, $\frac{1}{\sqrt{2}} = \frac{\sqrt{2}}{2}$ and $\frac{1}{\sqrt{3}} = \frac{\sqrt{3}}{3}$

The trigonometric ratios associated with the angles 30°, 45° and 60° (alongside their radian equivalents) are frequently used in problems involving trigonometry. Their **exact values** involve surds and can be easily obtained using either an equilateral triangle (of side two units) or an isosceles right-angled triangle.

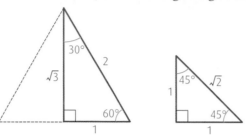

The trigonometric ratios for these triangles, together with those for angles of 0°, 90°, 180°, 270° and 360° (or their radian equivalent), are summarised in this table:

DON'T FORGET

You ought to know all of these values, as they are not given in the exam formula list. Knowing the shape of the trigonometric graphs will give you the ratios for 0°, 90°, 180°, 270° and 360°.

Radians	0	$\frac{\pi}{6}$	$\frac{\pi}{4}$	$\frac{\pi}{3}$	$\frac{\pi}{2}$	π	$\frac{3\pi}{2}$	2π
α	**0°**	**30°**	**45°**	**60°**	**90°**	**180°**	**270°**	**360°**
$\sin \alpha$	0	$\frac{1}{2}$	$\frac{\sqrt{2}}{2}$	$\frac{\sqrt{3}}{2}$	1	0	−1	0
$\cos \alpha$	1	$\frac{\sqrt{3}}{2}$	$\frac{\sqrt{2}}{2}$	$\frac{1}{2}$	0	−1	0	1
$\tan \alpha$	0	$\frac{1}{\sqrt{3}}$	1	$\sqrt{3}$	undefined	0	undefined	0

Example: 1

In triangle ABC, AB = $\sqrt{6}$ cm, AC = 3 cm and angle ACB = $\frac{\pi}{4}$.

Given that angle ABC is acute, calculate

(a) the size of angle ABC,

(b) the area of triangle ABC, giving your answer to 3 s.f.

DON'T FORGET

The sine rule and cosine rule from National 5 may have to be used from time to time in Higher, and the formulae are not given in the Higher examination formulae list.

DON'T FORGET

$\pi = 180°$, so the three angles in a triangle have a total of π radians.

Solution:

(a) Using the sine rule

$$\frac{\sin B}{b} = \frac{\sin C}{c} \Rightarrow \frac{\sin B}{3} = \frac{\sin\frac{\pi}{4}}{\sqrt{6}} \Rightarrow \frac{\sin B}{3} = \frac{\sqrt{2}}{2} \div \sqrt{6} = \frac{\sqrt{3}}{6}$$

$$\therefore \sin B = \frac{\sqrt{3}}{2} \Rightarrow \text{angle ABC} = \frac{\pi}{3}$$

We could say angle ABC = 60°, but for consistency with angle ACB in the question, we use $\frac{\pi}{3}$

(b) Area = $\frac{1}{2}bc\sin A$, where angle BAC = $\pi - \left(\frac{\pi}{4} + \frac{\pi}{3}\right) = \pi - \frac{7\pi}{12} = \frac{5\pi}{12}$

$$\therefore \text{Area} = \frac{1}{2} \times 3 \times \sqrt{6} \times \sin\left(\frac{5\pi}{12}\right) = 3.55 \text{ cm}^2 \text{ (3 s.f.)}$$

THE BASIC TRIGONOMETRIC GRAPHS

It is possible to sketch the graphs of the basic trigonometric functions, $y = \sin x$, $y = \cos x$ and $y = \tan x$ in terms of radians, as shown in the diagrams.

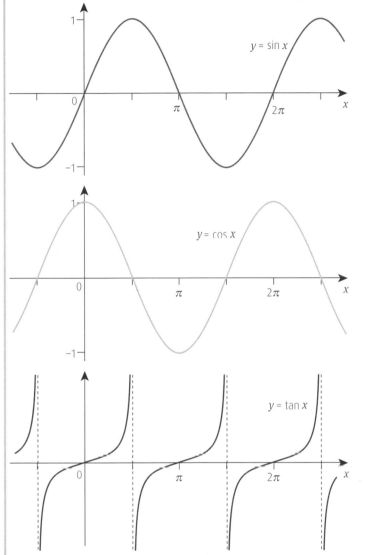

You should note the important features of these graphs.

The graphs of $y = \sin x$ and $y - \cos x$ are:
- **continuous** (i.e. there are no breaks)
- **periodic**, with period 2π (the period of a trigonometric graph is the number of degrees or radians for one cycle)
- contained entirely in the **amplitude range −1 to +1** i.e. $-1 \leqslant \sin x \leqslant 1$ and $-1 \leqslant \cos x \leqslant 1$. (The amplitude of each graph is 1; the graphs oscillate 1 unit above and 1 unit below the x-axis, with their maximum and minimum turning points occurring halfway between the zeros.)

The graph of $y = \tan x$:
- is **not continuous**. It is undefined when $x = ... -\frac{5\pi}{2}, -\frac{3\pi}{2}, -\frac{\pi}{2}, \frac{\pi}{2}, \frac{3\pi}{2}, \frac{5\pi}{2}$ ...etc.
- is **periodic**, with period π
- has an **unlimited amplitude range** i.e. $-\infty \leqslant \tan x \leqslant \infty$

ONLINE

For more on exact values and the trigonometric ratios for the four quadrants head to www.brightredbooks.net

ONLINE TEST

Head to www.brightredbooks.net to test yourself on the basics.

THINGS TO DO AND THINK ABOUT

1 What is the value of $\sin \frac{\pi}{3} - \cos \frac{5\pi}{4}$?

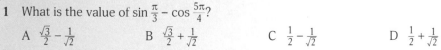

A $\frac{\sqrt{3}}{2} - \frac{1}{\sqrt{2}}$ B $\frac{\sqrt{3}}{2} + \frac{1}{\sqrt{2}}$ C $\frac{1}{2} - \frac{1}{\sqrt{2}}$ D $\frac{1}{2} + \frac{1}{\sqrt{2}}$ 2

TRIGONOMETRIC GRAPHS: PERIOD, AMPLITUDE AND GRAPH TRANSFORMATIONS 1

EXPRESSIONS AND FUNCTIONS

TRANSFORMATIONS OF TRIGONOMETRIC GRAPHS

The work in this section is essentially no different to that encountered in Chapter 1 of this book, when graphs of functions were **reflected** in either the x- or y-axis, **stretched** or **compressed** parallel to the x- or y-axis, and **translated** parallel to the x- or y-axis.

The table shows some examples of transformations on the basic trigonometric graphs, and the examples which follow make use of combinations of transformations.

$y = -\cos x$	The graph of $y = \cos x$ is reflected in the x-axis.	Reflections in the x-axis do not affect where the zeros occur.	
$y = \sin(-x)$	The graph of $y = \sin x$ is reflected in the y-axis.	Note that $y = \sin(-x) = -\sin x$, but that generally $y = f(-x) \neq -f(x)$	
$y = a\sin x$	The graph of $y = \sin x$ is stretched (or compressed) parallel to the y-axis (vertically) by a factor of a.	The amplitude of $y = a\sin x$ will be a. If a is negative, the graph will also be reflected in the x-axis.	

contd

$y = \tan bx$	The graph of $y - \tan x$ is compressed (or stretched) parallel to the x-axis (horizontally) by a factor of b.	The graph of $y = \tan 3x°$, $0 \le x \le 180$, is shown. The graph of $y = \tan x°$ is compressed so that three times as much is fitted into the same space. Hence, $y = \tan 3x°$ has three cycles in 180°, with each cycle taking $\frac{180°}{3} = 60°$ to complete.	
$y = \cos x + c$	The graph of $y = \cos x$ slides parallel to the y-axis: up ($c > 0$) or down ($c < 0$).	These vertical translations do not affect where the cycles of the graphs start and finish. They will all start together and take 2π (or 360°) for one cycle.	
$y = \sin(x - d)$	The graph of $y = \sin x$ slides parallel to the x-axis: to the right ($d > 0$) or to the left ($d < 0$).	The graph of $y = \sin(x - \frac{\pi}{4})$ is shown. The graph of $y = \sin x$ passes through $(0, 0)$ and $(\frac{\pi}{2}, 1)$. The graph of $y = \sin(x - \frac{\pi}{4})$ passes through $(\frac{\pi}{4}, 0)$ and $(\frac{3\pi}{4}, 1)$, so is a shift to the right of $\frac{\pi}{4}$.	

THINGS TO DO AND THINK ABOUT

1   The diagram shows the curve with equation of the form $y = \cos(x + a) + b$ for $0 \le x \le 2\pi$.

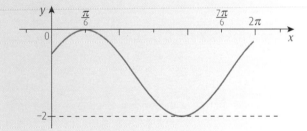

What is the equation of this curve?

A $y = \cos\left(x - \frac{\pi}{6}\right) - 1$

B $y = \cos\left(x - \frac{\pi}{6}\right) + 1$

C $y = \cos\left(x + \frac{\pi}{6}\right) - 1$

D $y = \cos\left(x + \frac{\pi}{6}\right) + 1$

2

ONLINE TEST

Test yourself on trigonometric graphs at www.brightredbooks.net

TRIGONOMETRIC GRAPHS: PERIOD, AMPLITUDE AND GRAPH TRANSFORMATIONS 2

EXPRESSIONS AND FUNCTIONS

TRANSFORMATIONS OF TRIGONOMETRIC GRAPHS

Example: 1

The graph with equation $y = a\sin(bx) + c$ is shown. Write down the values of a, b and c.

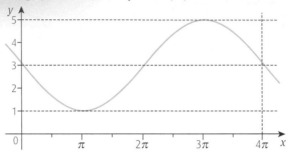

Solution:

Consider the value of b. The graph of $y = \sin x$ has a period of 2π, but this graph takes 4π for one cycle, so only half the amount fits into the usual space, meaning that $b = \frac{1}{2}$.

Next, consider a. This graph oscillates between 1 and 5, and therefore has an amplitude of 2 (i.e. it reaches 2 above and 2 below its midline). We know $y = \sin x$ moves from (0, 0) **upwards** to (π, 1) and this graph moves **downwards** from (0, 3) to (π, 1), so the value of $a = -2$.

Finally, consider the value of c. The midline of this graph is $y = 3$, whereas the midline of $y = \sin x$ is the x-axis ($y = 0$), so the graph has been moved vertically upwards by 3 units and $c = 3$.

The equation of this graph is therefore $y = -2\sin(\frac{1}{2}x) + 3$

Example: 2

Sketch the graph of $y = \cos x$, $0 \leq x \leq 2\pi$, and on the same diagram sketch the graph of $y = \frac{1}{2}\cos 3x - \frac{1}{2}$

Solution:

First sketch the graph of $y = \cos x$, but be aware that we will ultimately need to sketch a graph that has 3 cycles in 2π. It is a good idea, therefore, to mark the x-axis at least every $\frac{2\pi}{3}$ radians (intervals of $\frac{\pi}{3}$ would be even better).

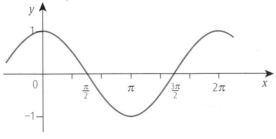

contd

Next, consider the graph of $y = \frac{1}{2}\cos 3x - \frac{1}{2}$ and compare it to $y = a\cos(bx) + c$, considering the order of operations.

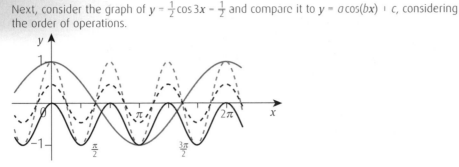

The graph must have 3 cycles in the usual space of 1 (since $b = 3$) i.e. each cycle will now take $\frac{2\pi}{3}$ to complete (think of this as transforming the graph of $y = \cos x$ to $y = \cos 3x$).

The amplitude of the graph is $\frac{1}{2}$ ($a = \frac{1}{2}$), so the graph must oscillate between $\frac{1}{2}$ above and $\frac{1}{2}$ below its midline (think of this as transforming the graph of $y = \cos 3x$ to $y = \frac{1}{2}\cos 3x$).

The graph has been shifted vertically downwards by $\frac{1}{2}$ unit ($c = -\frac{1}{2}$), so its midline is $y = -\frac{1}{2}$ (think of this as transforming the graph of $y = \frac{1}{2}\cos 3x$ to $y = \frac{1}{2}\cos 3x - \frac{1}{2}$).

THINGS TO DO AND THINK ABOUT

1 Find the maximum value of $2 - 3\sin\left(x - \frac{\pi}{3}\right)$

and the value of x where this occurs in the interval $0 \leqslant x \leqslant 2\pi$.

	max value	x
A	−1	$\frac{11\pi}{6}$
B	5	$\frac{11\pi}{6}$
C	−1	$\frac{5\pi}{6}$
D	5	$\frac{5\pi}{6}$

2

2 Which of the following shows the graph of $y = 4\cos 2x - 1$, for $0 \leqslant x \leqslant \pi$?

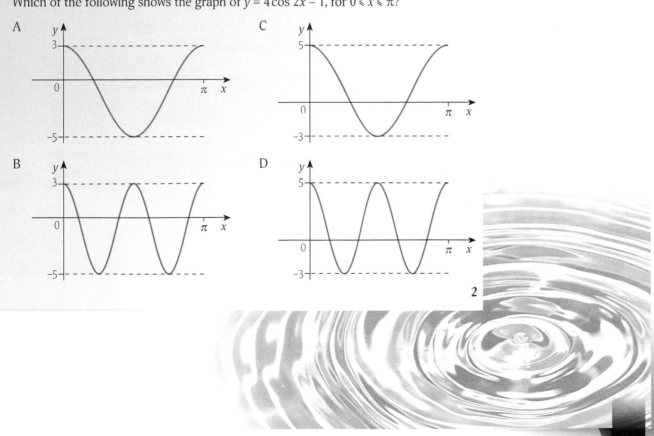

2

THE ADDITION AND DOUBLE-ANGLE FORMULAE ⟨ EXPRESSIONS AND FUNCTIONS ⟩

THE ADDITION FORMULAE

You can use the **addition formulae** to find the sine or cosine of the sum or difference of two angles.

When you have an expression like $\sin(x + 30°)$ or $\cos\left(t - \frac{\pi}{3}\right)$, you can use the following formulae to expand the brackets:

$$\sin(A \pm B) = \sin A \cos B \pm \cos A \sin B$$
$$\cos(A \pm B) = \cos A \cos B \mp \sin A \sin B$$
(These are given in the Higher exam formulae list.)

Example: 1

Given that $\cos(A - B) = \cos A \cos B + \sin A \sin B$, show that $\cos 15°$ can be written in the form $\frac{\sqrt{a} + \sqrt{b}}{c}$, stating the values of a, b and c.

Solution:

This is a non-calculator question, so you will be expected to make use of the exact values from earlier in this chapter within your answer.

The key is to find a pair of angles which have a difference of 15°, and to substitute these into the addition formula.

$$\cos 15° = \cos(45° - 30°)$$
$$= \cos 45° \cos 30° + \sin 45° \sin 30°$$
$$= \left(\frac{\sqrt{2}}{2}\right)\left(\frac{\sqrt{3}}{2}\right) + \left(\frac{\sqrt{2}}{2}\right)\left(\frac{1}{2}\right)$$
$$= \frac{\sqrt{6}}{4} + \frac{\sqrt{2}}{4}$$
$$= \frac{\sqrt{6} + \sqrt{2}}{4}$$
$$\therefore a = 6, b = 2, c = 4$$

Example: 2

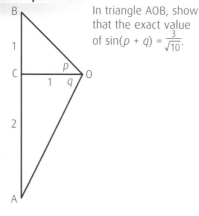

In triangle AOB, show that the exact value of $\sin(p + q) = \frac{3}{\sqrt{10}}$.

Solution:

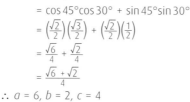
expand $\sin(p + q)$ using the addition formula.

$$\sin(p + q) = \sin p \cos q + \cos p \sin q$$

$\sin p = \frac{BC}{OB}$ and $\cos p = \frac{OC}{OB}$, so we need to find the length OB.

$OB = \sqrt{(1^2 + 1^2)} = \sqrt{2}$. Similarly, $OA = \sqrt{5}$.

substitute the values of the 4 ratios into the formula.

$$\sin(p + q) = \sin p \cos q + \cos p \sin q$$
$$= \left(\frac{1}{\sqrt{2}}\right)\left(\frac{1}{\sqrt{5}}\right) + \left(\frac{1}{\sqrt{2}}\right)\left(\frac{2}{\sqrt{5}}\right)$$
$$= \frac{1}{\sqrt{10}} + \frac{2}{\sqrt{10}}$$
$$= \frac{3}{\sqrt{10}} \text{ as required.}$$

THE DOUBLE-ANGLE FORMULAE

Double-angle formulae can be used to turn any trigonometric expression containing a '$2x$'-type term into one containing 'single x' terms, and are often used in questions involving either trigonometric identities or the solution of trigonometric equations.

contd

$$\sin 2A = 2\sin A \cos A$$
$$\cos 2A = \cos^2 A - \sin^2 A$$
$$\qquad = 2\cos^2 A - 1$$
$$\qquad = 1 - 2\sin^2 A$$

(These are given in the Higher exam formulae list.)

Example: 3

By making use of the addition formula for cos(A + B), show that $\cos 3x = 4\cos^3 x - 3\cos x$.

Solution:

Firstly, write $\cos 3x$ as $\cos(2x + x)$ in order to use the addition formula.

$$\cos 3x = \cos(2x + x) = \cos 2x \cos x - \sin 2x \sin x$$

Now, note that the result we are trying to prove does not contain '2x', so make use of the double-angle formulae.

$$\cos 2x \cos x - \sin 2x \sin x = (2\cos^2 x - 1)\cos x - (2\sin x \cos x)\sin x$$
$$= 2\cos^3 x - \cos x - 2\sin^2 x \cos x$$
$$= 2\cos^3 x - \cos x - 2(1 - \cos^2 x)\cos x$$
$$= 2\cos^3 x - \cos x - 2(\cos x - \cos^3 x)$$
$$= 2\cos^3 x - \cos x - 2\cos x + 2\cos^3 x \qquad = 4\cos^3 x - 3\cos x \text{ as required.}$$

> **DON'T FORGET**
>
> $\cos 2x$ has three alternatives, but we want the one containing $\cos x$ only

> **DON'T FORGET**
>
> You may require prior knowledge from National 5, such as the identity $\sin^2 A + \cos^2 A \equiv 1$

Example: 4

In triangle ABC, AB = 4 cm, BC = 1 cm, angle ABC = 90° and angle BAC = θ.

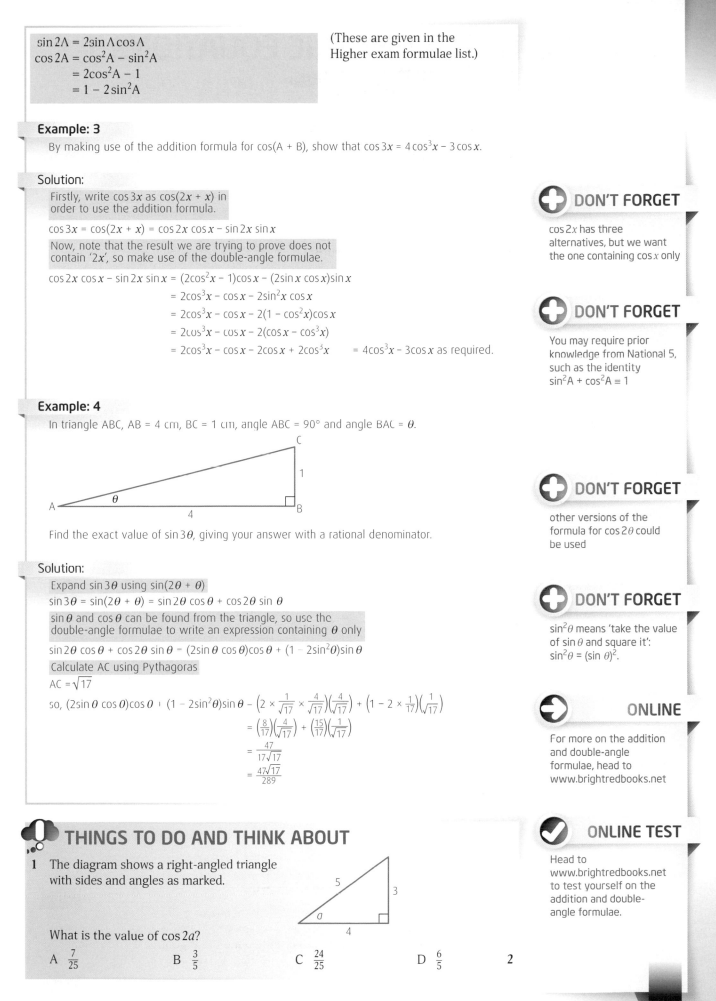

Find the exact value of $\sin 3\theta$, giving your answer with a rational denominator.

Solution:

Expand $\sin 3\theta$ using $\sin(2\theta + \theta)$

$$\sin 3\theta = \sin(2\theta + \theta) = \sin 2\theta \cos \theta + \cos 2\theta \sin \theta$$

$\sin \theta$ and $\cos \theta$ can be found from the triangle, so use the double-angle formulae to write an expression containing θ only

$$\sin 2\theta \cos \theta + \cos 2\theta \sin \theta = (2\sin \theta \cos \theta)\cos \theta + (1 - 2\sin^2 \theta)\sin \theta$$

Calculate AC using Pythagoras

$$AC = \sqrt{17}$$

So, $(2\sin\theta\cos\theta)\cos\theta + (1 - 2\sin^2\theta)\sin\theta = \left(2 \times \frac{1}{\sqrt{17}} \times \frac{4}{\sqrt{17}}\right)\left(\frac{4}{\sqrt{17}}\right) + \left(1 - 2 \times \frac{1}{17}\right)\left(\frac{1}{\sqrt{17}}\right)$

$$= \left(\frac{8}{17}\right)\left(\frac{4}{\sqrt{17}}\right) + \left(\frac{15}{17}\right)\left(\frac{1}{\sqrt{17}}\right)$$

$$= \frac{47}{17\sqrt{17}}$$

$$= \frac{47\sqrt{17}}{289}$$

> **DON'T FORGET**
>
> other versions of the formula for $\cos 2\theta$ could be used

> **DON'T FORGET**
>
> $\sin^2\theta$ means 'take the value of $\sin\theta$ and square it': $\sin^2\theta = (\sin\theta)^2$.

> **ONLINE**
>
> For more on the addition and double-angle formulae, head to www.brightredbooks.net

THINGS TO DO AND THINK ABOUT

1. The diagram shows a right-angled triangle with sides and angles as marked.

What is the value of $\cos 2a$?

A $\frac{7}{25}$ B $\frac{3}{5}$ C $\frac{24}{25}$ D $\frac{6}{5}$ 2

> **ONLINE TEST**
>
> Head to www.brightredbooks.net to test yourself on the addition and double-angle formulae.

SOLVING TRIGONOMETRIC EQUATIONS 1
RELATIONSHIPS AND CALCULUS

BASIC EQUATIONS IN A GIVEN INTERVAL

At National 5 you were required to solve trigonometric equations which reduce down to those of the form $\sin x = k$, $\cos x = k$ or $\tan x = k$. There are two ways to find solutions in an interval:

- use the known symmetry and periodicity of the graph of the functions

- use the quadrant grid (CAST).

Example: 1

Solve $\sin x = -\frac{3}{4}$ for $0 \leq x \leq 2\pi$

Solution:

Method 1

Draw the graphs of $y = \sin x$ and and $y = -\frac{3}{4}$ on the same axes (you may have to go beyond the range you are interested in in order to help here).

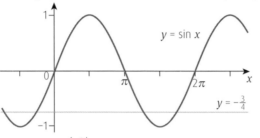

Calculate $\sin^{-1}\left(-\frac{3}{4}\right)$ on your calculator (this gives –0·848 to 3 s.f.) and use the symmetry and periodicity of the graph to calculate the solutions.

$\therefore x = \pi + 0\cdot848 = 3\cdot99$ (3 s.f.) or $x = 2\pi - 0\cdot848 = 5\cdot44$ (3 s.f.)

Method 2

Use the quadrant grid (CAST diagram) .

Firstly, ignoring the negative find the solution of $\sin x = \frac{3}{4}$ from your calculator, using $\sin^{-1}\left(\frac{3}{4}\right) = 0\cdot848$

Mark this angle on the CAST diagram, in the 'A' quadrant, alongside the other three related angles in the S, T and C quadrants.

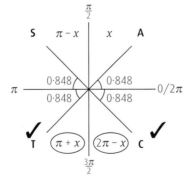

Now select those quadrants where $\sin x$ is negative $\left(\text{since } \sin x = -\frac{3}{4}\right)$ i.e. 'T' and 'C'.

$\therefore x = \pi + 0\cdot848 = 3\cdot99$ (3 s.f.) or $x = 2\pi - 0\cdot848 = 5\cdot44$ (3 s.f.)

DON'T FORGET

An angle measured in radians does not necessarily have to be expressed in terms of π.

contd

Example: 2

Solve cos $(2x - 15)° = 0.11$ for $0 \leq x \leq 360$, giving your answers to 1 decimal place.

Solution:

Since $0 \leq 2x \leq 720$ and $-15 \leq 2x - 15 \leq 705$, we need to complete two revolutions of the CAST diagram to ensure we find all of the solutions asked for.

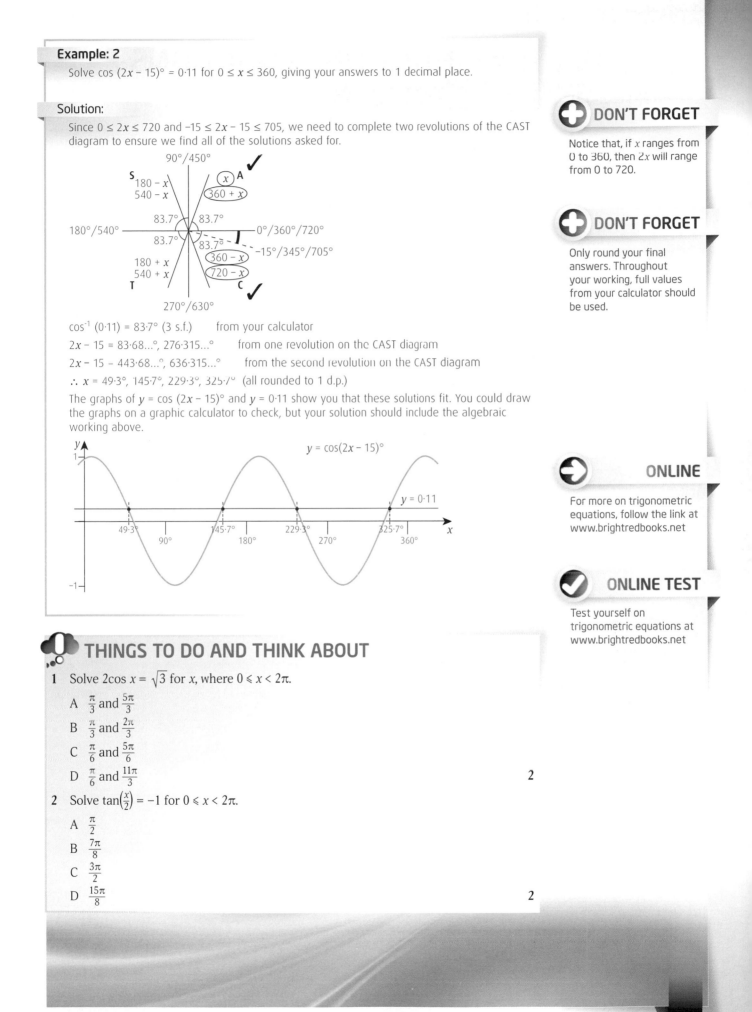

$\cos^{-1}(0.11) = 83.7°$ (3 s.f.)   from your calculator

$2x - 15 = 83.68...°, 276.315...°$ from one revolution on the CAST diagram

$2x - 15 = 443.68...°, 636.315...°$ from the second revolution on the CAST diagram

$\therefore x = 49.3°, 145.7°, 229.3°, 325.7°$ (all rounded to 1 d.p.)

The graphs of $y = \cos (2x - 15)°$ and $y = 0.11$ show you that these solutions fit. You could draw the graphs on a graphic calculator to check, but your solution should include the algebraic working above.

DON'T FORGET

Notice that, if x ranges from 0 to 360, then $2x$ will range from 0 to 720.

DON'T FORGET

Only round your final answers. Throughout your working, full values from your calculator should be used.

ONLINE

For more on trigonometric equations, follow the link at www.brightredbooks.net

ONLINE TEST

Test yourself on trigonometric equations at www.brightredbooks.net

THINGS TO DO AND THINK ABOUT

1 Solve $2\cos x = \sqrt{3}$ for x, where $0 \leqslant x < 2\pi$.

   A $\frac{\pi}{3}$ and $\frac{5\pi}{3}$

   B $\frac{\pi}{3}$ and $\frac{2\pi}{3}$

   C $\frac{\pi}{6}$ and $\frac{5\pi}{6}$

   D $\frac{\pi}{6}$ and $\frac{11\pi}{3}$ 2

2 Solve $\tan\left(\frac{x}{2}\right) = -1$ for $0 \leqslant x < 2\pi$.

   A $\frac{\pi}{2}$

   B $\frac{7\pi}{8}$

   C $\frac{3\pi}{2}$

   D $\frac{15\pi}{8}$ 2

SOLVING TRIGONOMETRIC EQUATIONS 2
RELATIONSHIPS AND CALCULUS

SOLVING EQUATIONS USING THE DOUBLE-ANGLE FORMULAE

The key thing to note with equations which involve the double-angle formulae is that angles such as 'x' and '$2x$' will appear together in the one equation. These equations will involve:

$\sin 2x$ and either $\sin x$ or $\cos x$ $\cos 2x$ and $\cos x$ $\cos 2x$ and $\sin x$

Example: 1

Solve $3 \sin 2x + 5 \sin x = 0$, $0° \le x° < 360°$

Solution:

The only substitution possible is $\sin 2x \equiv 2 \sin x \cos x$, so the equation becomes:

$3(2 \sin x \cos x) + 5 \sin x = 0$, $0° \le x° < 360°$

$6 \sin x \cos x + 5 \sin x = 0$

$\sin x(6 \cos x + 5) = 0$

$\sin x = 0$, or $6 \cos x + 5 = 0$

$\cos x = -\frac{5}{6}$

> You could always use 's' and 'c' for shorthand if you like when solving these equations (e.g. $6sc + 5s = 0$), but at this stage you **must** write $\sin x = 0$ and $\cos x = -\frac{5}{6}$, not $s = 0$ and $c = -\frac{5}{6}$.

DON'T FORGET

Consider carefully the interval in question. 360° is not a solution here as x must be strictly less than 360°.

The solutions for $\sin x° = 0$ can be found from the graph of $y = \sin x°$.

Since $\cos^{-1}\left(\frac{5}{6}\right) = 33·557...°$, $\cos x° = -\frac{5}{6}$ when $x = 180 - 33·557...°$ and $180 + 33·557...°$.

$\therefore x = 0°, 180°$, or $x = 146·4°, 213·6°$

Example: 2

Solve $4 - 7 \sin \theta - \cos 2\theta = 0$, $0 \le \theta \le 2\pi$

Solution:

This is a non-calculator question, so you will be expected to make use of the exact values from earlier in this chapter within your answer.

Firstly, use a substitution for $\cos 2\theta$ to create an equation involving just θ.

Here, since our equation already involves $\sin \theta$, we use $\cos 2\theta \equiv 1 - 2\sin^2\theta$ so we don't end up with a mixture of sine and cosine terms:

$4 - 7 \sin \theta - (1 - 2\sin^2\theta) = 0$, $0 \le \theta \le 2\pi$

$4 - 7 \sin \theta - 1 + 2\sin^2\theta = 0$

$2 \sin^2\theta - 7 \sin \theta + 3 = 0$

Now solve this quadratic equation in θ using either factorisation or the quadratic formula:

$(2 \sin \theta - 1)(\sin \theta - 3) = 0$

> You could use $2s^2 - 7s + 3 = 0$ here, leading to $(2s - 1)(s - 3) = 0$

$\sin \theta = \frac{1}{2}$ or $\sin \theta = 3$, no real solutions, since $-1 \le \sin \theta \le 1$

$\sin^{-1}\left(\frac{1}{2}\right) = \frac{\pi}{6}$, so $\sin \theta = \frac{1}{2}$ when $\theta = \frac{\pi}{6}$ and $\pi - \frac{\pi}{6}$.

$\therefore x = \frac{\pi}{6}, \frac{5\pi}{6}$

contd

Example: 3

Solve $2\cos 4t + 1 = 3\cos 2t$, $0 \le t \le \pi$, giving your answers to 2 decimal places.

Solution:

$4t$ is twice $2t$, so make the substitution $\cos 4t = 2\cos^2 2t - 1$

$2(2\cos^2 2t - 1) + 1 = 3\cos 2t$, $0 \le t \le \pi$ an equation involving $\cos 2t$ only

$4\cos^2 2t - 2 + 1 = 3\cos 2t$

$4\cos^2 2t - 3\cos 2t - 1 = 0$ equate to zero to create quadratic

$(\cos 2t - 1)(4\cos 2t + 1) = 0$

$\cos 2t = 1$ or $\cos 2t = \frac{1}{4}$ factorise or use the quadratic formula

The solutions for $\cos 2t = 1$ can be found with relative ease from the graph of $y = \cos 2t$, $0 \le t \le \pi$.

Since $\cos^{-1}\left(\frac{1}{4}\right) = 1\cdot318...$, $\cos 2t = -\frac{1}{4}$ when $2t = \pi - 1\cdot318...$, $\pi + 1\cdot318...$

$\therefore\ t = 0, \pi$ or $t = 0\cdot91, 2\cdot23$ (2 d.p.)

DON'T FORGET

$0 \le t \le \pi$ leads to $0 \le 2t \le 2\pi$, so solutions for $\cos 2t$ must be considered from one revolution of the CAST diagram here.

THINGS TO DO AND THINK ABOUT

1 (a) Solve $\cos 2x° - 3\cos x° + 2 = 0$ for $0 \le x < 360$.   5

   (b) Hence solve $\cos 4x° - 3\cos 2x° + 2 = 0$ for $0 \le x < 360$.   2

2 Solve algebraically the equation $\sin 2x = 2\cos^2 x$ for $0 \le x < 2\pi$.   6

3 Solve the equation $\sin x - 2\cos 2x = 1$ for $0 \le x < 2\pi$.   5

ONLINE

For more on trigonometric equations, follow the link at www.brightredbooks.net

ONLINE TEST

Test yourself on trigonometric equations at www.brightredbooks.net

THE WAVE FUNCTION 1
EXPRESSIONS AND FUNCTIONS

EXPRESSING $p\sin x \pm q\cos x$ AS A SINGLE TRIGONOMETRIC FUNCTION

When two waves are added together a new wave is created – you can see this at the beach or the swimming pool. The sum or difference between a sine and a cosine wave can be expressed as a new single trigonometric function by making use of addition formulae.

The waves you work with will have the same period, so they are both functions of the same variable: x, or 2θ, or $0{\cdot}5t$, for example, and the combined wave will have the same period, but with a **phase shift**.

Example: 1

Show clearly that $p\sin x + q\cos x$ can be written in the form $k\cos(x - a)$, where $k = \sqrt{p^2 + q^2}$ and $a = \tan^{-1}\left(\frac{p}{q}\right)$

Solution:

Using addition formulae, we can write $k\cos(x - a)$ as $k(\cos x\cos a + \sin x\sin a)$.

So, equating $p\sin x + q\cos x$ with $k\cos(x - a)$, and expanding the bracket on the right-hand side by k gives:

$p\sin x + q\cos x = k\cos(x - a)$

$p\sin x + q\cos x = k(\cos x\cos a + \sin x\sin a)$

$p\sin x + q\cos x = k\cos x\cos a + k\sin x\sin a$

$\boldsymbol{p}\sin x + \boldsymbol{q}\cos x = \boldsymbol{k\sin a}\sin x + \boldsymbol{k\cos a}\cos x$ rearranging here allows for easier comparison

For the two sides to be equal for all values of x, the coefficient of $\sin x$ on the left-hand side must equal the coefficient of $\sin x$ on the right-hand side, and similarly for the coefficients of $\cos x$.

We can see, therefore, in **red** that $\boldsymbol{p} = \boldsymbol{k\sin a}$ and in **blue** that $\boldsymbol{q} = \boldsymbol{k\cos a}$

Squaring both equations and adding gives

$p^2 + q^2 = k^2\sin^2 a + k^2\cos^2 a$

$p^2 + q^2 = k^2(\sin^2 a + \cos^2 a)$

$p^2 + q^2 = k^2$ since $\sin^2 a + \cos^2 a \equiv 1$

$\therefore k = \sqrt{p^2 + q^2}$

Also $\frac{k\sin a}{k\cos a} = \frac{p}{q} \Rightarrow \tan a = \frac{p}{q}$

$\therefore a = \tan^{-1}\left(\frac{p}{q}\right)$ as required

Example: 2

Express $5\cos x - 12\sin x$ in the form $k\sin(x + a)$, where $k > 0$ and $0 \leq a \leq 2\pi$.

Solution:

$5\cos x - 12\sin x = k\sin(x + a)$

$\qquad\qquad\qquad = k(\sin x\cos a + \cos x\sin a)$

$\qquad\qquad\qquad = k\sin x\cos a + k\cos x\sin a$

$\qquad\qquad\qquad = \boldsymbol{k\sin a}\cos x + \boldsymbol{k\cos a}\sin x$

$k\sin a = 5$

$k\cos a = -12$

$k = \sqrt{5^2 + (-12)^2} = \sqrt{25 + 144} = 13$ $k > 0$, so negative root not considered

$\frac{k\sin a}{k\cos a} = \frac{5}{-12} \Rightarrow \tan a = -\frac{5}{12}$

Since $k > 0$, angle a is such that $\sin a$ is positive, $\cos a$ is negative and $\tan a$ is negative.

We need the angle a to be in the correct quadrant to satisfy this – and that means the 'S' quadrant.

$\tan^{-1}\left(\frac{5}{12}\right) = 0{\cdot}39... \Rightarrow a = \pi - 0{\cdot}39... = 2{\cdot}7468...$

$\therefore 5\cos x - 12\sin x = 13\sin(x + 2{\cdot}75)$

DON'T FORGET

This is the quick way to find k and is acceptable in the exam without the need to formally use $\sin^2 a + \cos^2 a \equiv 1$

USING THE WAVE FUNCTION TO SKETCH GRAPHS

Rewriting an expression of the form $p\sin x + q\cos x$ as a single trigonometric function can help us sketch the graph more easily – it will be more obvious what the amplitude and phase shift are.

Example: 3

(a) Express $\cos x - \sqrt{3}\sin x$ in the form $k\sin(x - a)$, where $k > 0$ and $0 \le a \le 2\pi$.

(b) Hence, or otherwise, sketch the curve with equation $y = 3 + \cos x - \sqrt{3}\sin x$, $0 \le x \le 2\pi$.

Solution:

(a) $\cos x - \sqrt{3}\sin x = k\sin(x - a)$

$\qquad\qquad = k(\sin x \cos a - \cos x \sin a)$

$\qquad\qquad = k\sin x \cos a - k\cos x \sin a$

$\qquad\qquad = \mathbf{-k\sin a \cos x + k\cos a \sin x}$

$\mathbf{-k\sin a = 1} \to \mathbf{k\sin a = -1}$ $\qquad$ careful with the negatives at this stage

$\mathbf{k\cos a = -\sqrt{3}}$

$k = \sqrt{(-1)^2 + (-\sqrt{3})^2} = \sqrt{1 + 3} = 2$

$\frac{k\sin a}{k\cos a} = \frac{-1}{-\sqrt{3}} \to \tan a = \frac{1}{\sqrt{3}}$

$\sin a$ is $-$ve, $\cos a$ is $-$ve, $\tan a$ is $+$ve, so a lies in 'T' quadrant.

$a = \pi + \frac{\pi}{6} = \frac{7\pi}{6}$

$\therefore \cos x - \sqrt{3}\sin x = 2\sin\left(x - \frac{7\pi}{6}\right)$

(b) Using part (a), $y = 3 + \cos x - \sqrt{3}\sin x$ can be rewritten as $y = 3 + 2\sin\left(x - \frac{7\pi}{6}\right)$.

Hence, we need to sketch the graph of $y = 2\sin\left(x - \frac{7\pi}{6}\right) + 3$, which is a transformation of $y = \sin x$.

$y = \sin x \to y = \sin\left(x - \frac{7\pi}{6}\right)$

a horizontal translation by $\frac{7\pi}{6}$ to the right
maximum at $\left(\frac{\pi}{2}, 1\right)$ now at $\left(\frac{\pi}{2} + \frac{7\pi}{6}, 1\right)$ i.e. $\left(\frac{5\pi}{3}, 1\right)$
minimum at $\left(\frac{-\pi}{2}, -1\right)$ now at $\left(\frac{-\pi}{2} + \frac{7\pi}{6}, -1\right)$ i.e. $\left(\frac{2\pi}{3}, -1\right)$
zeros at $\left(-\pi + \frac{7\pi}{6}, 0\right)$ and $\left(0 + \frac{7\pi}{6}, 0\right)$ i.e. $\left(\frac{\pi}{6}, 0\right)$ and $\left(\frac{7\pi}{6}, 0\right)$

$y = \sin\left(x - \frac{7\pi}{6}\right) \to y = 2\sin\left(x - \frac{7\pi}{6}\right)$

amplitude doubled from 1 to 2
maximum at $\left(\frac{5\pi}{3}, 2\right)$
minimum at $\left(\frac{2\pi}{3}, -2\right)$
zeros remain at $\left(\frac{\pi}{6}, 0\right)$ and $\left(\frac{7\pi}{6}, 0\right)$

$y = 2\sin\left(x - \frac{7\pi}{6}\right) \to y = 2\sin\left(x - \frac{7\pi}{6}\right) + 3$

vertical translation upwards by 3 units
maximum at $\left(\frac{5\pi}{3}, 5\right)$
minimum at $\left(\frac{2\pi}{3}, 1\right)$
$\left(\frac{\pi}{6}, 0\right) \to \left(\frac{\pi}{6}, 3\right)$ and $\left(\frac{7\pi}{6}, 0\right) \to \left(\frac{7\pi}{6}, 3\right)$

The graph of $y = 3 + \cos x - \sqrt{3}\sin x$ crosses the y-axis when $x = 0$

$\therefore y = 3 + \cos 0 - \sqrt{3}\sin 0 = 3 + 1 - 0 = 4$

i.e. at the point $(0, 4)$

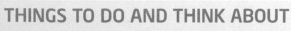

DON'T FORGET

"Hence, or otherwise" should be taken to mean "we strongly recommend you do this by using the result we asked you to find earlier, but if you really want to you can do it some other way"!

DON'T FORGET

The graph must only cover the range of values required; in this case, $0 \le x \le 2\pi$. The y-coordinate when $x = 0$ should be equal to the y-coordinate when $x = 2\pi$, one wavelength later.

DON'T FORGET

The points on the graph's midline (when $y = 3$) occur halfway between the maximum and minimum turning points.

ONLINE

For more on the wave function, including its applications, follow the links at www.brightredbooks.net

THINGS TO DO AND THINK ABOUT

1 (a) The expression $3\sin x - 5\cos x$ can be written in the form $R\sin(x + a)$ where $R > 0$ and $0 \le a < 2\pi$. Calculate the values of R and a. **4**

(b) Hence find the value of t, where $0 \le t \le 2$, for which $\int_0^t (3\cos x + 5\sin x)dx = 3$. **7**

ONLINE TEST

Test yourself on the wave function at www.brightredbooks.net

THE WAVE FUNCTION 2
EXPRESSIONS AND FUNCTIONS

USING THE WAVE FUNCTION TO SOLVE EQUATIONS

Example: 1

The function $f(\theta) = \sqrt{2}\sin\theta - 3\cos\theta$.

Find the maximum value of $f(\theta)$ and the value of θ where this occurs, given that $0 \leq \theta \leq 2\pi$.

Solution:

Notice that in this example you are not told which wave form to use (in fact, you are not even told to use the wave function at all...you are expected to recognise the need for it here).

In this case, any of the four available options will work, but it is worth spending some time, before answering the question, to consider which of the wave forms will make life easier for you.

Since the expression $\sqrt{2}\sin\theta - 3\cos\theta$ contains a '−' sign, the best choices will be either $k\cos(\theta + a)$ or $k\sin(\theta - a)$, since expanding these using the addition formulae results in expressions containing a '−'.

$k\cos(\theta + a) = k\cos a\cos\theta - k\sin a\sin\theta$ leads to both $k\cos a$ and $k\sin a$ being negative... certainly not impossible to work with, but care is needed surrounding negatives and the correct quadrant for the value of a!

$k\sin(\theta - a) = k\cos a\sin\theta - k\sin a\cos\theta$ leads to both $k\cos a$ and $k\sin a$ being positive... a nice option.

$$\sqrt{2}\sin\theta - 3\cos\theta = k\sin(\theta - a)$$
$$= k(\sin\theta\cos a - \cos\theta\sin a)$$
$$= k\sin\theta\cos a - k\cos\theta\sin a$$
$$= \boldsymbol{k\cos a}\sin\theta - \boldsymbol{k\sin a}\cos\theta$$

$k\cos a = \sqrt{2}$

$k\sin a = 3$

$k = \sqrt{(\sqrt{2})^2 + 3^2} = \sqrt{11}$

$\dfrac{k\sin a}{k\cos a} = \dfrac{3}{\sqrt{2}} \rightarrow \tan a = \dfrac{3}{\sqrt{2}}$

$\sin a$ +ve, $\cos a$ +ve, $\tan a$ +ve $\rightarrow a$ lies in 'A' quadrant.

$\therefore \sqrt{2}\sin\theta - 3\cos\theta = \sqrt{11}\sin(\theta - 1 \cdot 13)$

Since $\sqrt{11}\sin(\theta - 1 \cdot 13)$ oscillates between $\sqrt{11}$ and $-\sqrt{11}$, the máximum value of $f(\theta)$ is $\sqrt{11}$.

This maximum occurs when $\sqrt{11}\sin(\theta - 1 \cdot 13) = \sqrt{11}$

i.e. $\sin(\theta - 1 \cdot 13) = 1$

$\therefore \theta - 1 \cdot 13 = \dfrac{\pi}{2}$

$\theta = \dfrac{\pi}{2} + 1 \cdot 13 = 2 \cdot 70$ (3 s.f.)

DON'T FORGET

When no specific wave form is asked for, any of the four will work, but some may make life easier than others!

DON'T FORGET

This example could also be solved using calculus – differentiating to find an expresion for $f'(\theta)$ and solving $f'(\theta) = 0$. Remember to then justify that $\theta = 2 \cdot 70$ is a maximum value using either $f'(\theta)$ or a nature table.

Example: 2

The formula $d(t) = 250(\cos 30t - \sin 30t) + 450$ gives the depth, in cm, of water in a harbour t hours after midnight.

(a) Express $\cos 30t - \sin 30t$ in the form $k\cos(30t + a)$, where $k > 0$ and $0° \leq a° \leq 360°$.

(b) Hence, or otherwise, find the depth of water in the harbour at high tide (when the water is at its maximum depth) and at what time this first occurs.

(c) A vessel arrives at 4am and needs a 3 metre depth to sail into the harbour. How long must it wait until it can safely enter the harbour?

Solution:

(a) $\cos 30t - \sin 30t = k\cos(30t + a)$
$$= k(\cos 30t\cos a - \sin 30t\sin a)$$
$$= k\cos 30t\cos a - k\sin 30t\sin a$$
$$= \boldsymbol{k\cos a}\cos 30t - \boldsymbol{k\sin a}\sin 30t$$

$k\sin a = 1$

$k\cos a = 1$

contd

$k = \sqrt{1^2 + 1^2} = \sqrt{2}$

$\frac{k\sin a}{k\cos a} = \frac{1}{1} \rightarrow \tan a = 1$

$\sin a$ +ve, $\cos a$ +ve, $\tan a$ +ve $\rightarrow a$ lies in 'A' quadrant.

$\therefore \cos 30t - \sin 30t = \sqrt{2}\cos(30t + 45)$

(b) You can sketch the graph of $d(t)$

$= 250(\cos 30t - \sin 30t) + 450$

$= 250\sqrt{2}\cos(30t + 45) + 450$

if you wish to, though it is not essential in this question.

The most important thing to understand in this part of the question is that the graph of $d(t) = 250\sqrt{2}\cos(30t + 45) + 450$ will have a maximum value of $250\sqrt{2} + 450$ (since $\cos(30t + 45)$ has a maximum value of 1).

This maximum value occurs when $\cos(30t + 45) = 1$

$30t + 45 = 0, 360, 720, ...$

$30t = -45, 315, 675, ...$

$t = 10\cdot5, ...$ (we don't need more values)

$\therefore$ At high tide, the depth of water in the harbour is $250\sqrt{2} + 450 \approx 803\cdot55$cm

High tide first occurs 10·5 hours after midnight i.e. at 10:30am.

(c) We need to know when the depth of water in the harbour is at 3 metres i.e. 300cm

$250\sqrt{2}\cos(30t + 45) + 450 = 300$

$250\sqrt{2}\cos(30t + 45) = -150$

$\cos(30t + 45) = -0\cdot424...$

Since $\cos^{-1}(0\cdot424...) \approx 64\cdot9°$, $\cos(30t° + 45) = -0\cdot424...$ when $30t + 45 = 180 - 64\cdot9$ or $30t + 45 = 180 + 64\cdot9$

so $t = 2\cdot34$ or $t = 6\cdot66$ (3 s.f.)

Since the vessel arrived at 4am, i.e. $t = 4$, it must wait for $(6\cdot66 - 4)$ hours, i.e. approximately 2 h 40 min (until 6:40am)

$d(t) = 250\sqrt{2}\cos(30t + 45)° + 450$

THINGS TO DO AND THINK ABOUT

1 (a) The expression $\cos x - \sqrt{3}\sin x$ can be written in the form $k\cos(x + a)$ where $k > 0$ and $0 \leqslant a < 2\pi$.
Calculate the values of k and a. **4**

(b) Find the points of intersection of the graph of $y = \cos x - \sqrt{3}\sin x$ with the x and y axes, in the interval $0 \leqslant x \leqslant 2\pi$. **3**

2 (a) The expression $\sqrt{3}\sin x° - \cos x°$ can be written in the form $k\sin(x - a)°$, where $k > 0$ and $0 \leqslant a < 360$.
Calculate the values of k and a. **4**

(b) Determine the maximum value of $4 + 5\cos x° - 5\sqrt{3}\sin x°$, where $0 \leqslant x < 360$. **2**

3 If $3\sin x - 4\cos x$ is written in the form $k\cos(x - a)$, what are the values of $k\cos a$ and $k\sin a$?

	$k\cos a$	$k\sin a$
A	-3	4
B	3	-4
C	4	-3
D	-4	3

2

ONLINE

For more on the wave function, including its applications, follow the links at www.brightredbooks.net

ONLINE TEST

Test yourself on the wave function at www.brightredbooks.net

SYLLABUS SKILLS CHECKLISTS

ALGEBRA

Algebra	
A1	Know and use the terms **range** and **domain**. ☐
A2	Recognise general features of **polynomial**, **exponential** and **logarithmic** graphs. ☐
A3	**Determine function** (polynomial, exponential, logarithmic) from graph and vice versa. ☐
A4	Identify or sketch a function after a **transformation** of the form $kf(x)$, $f(kx)$, $f(x + k)$, $f(x) + k$ or a combination of these (k is real). ☐
A5	Determine and sketch $f^{-1}(x)$ (**inverse**) of functions where it exists. ☐
A6	Given the graph of $y = f(x)$, **sketch $f'(x)$** and vice versa. ☐
A7	Determine a **composite function** given $f(x)$ and $g(x)$, where $f(x)$ and $g(x)$ can be trigonometric, logarithmic, exponential or algebraic functions. ☐
A8	**Complete the square** in a quadratic expression where the coefficient of x^2 is non-unitary. ☐
A9	Solve a **quadratic inequality**. ☐
A10	Find the **nature of roots** of a quadratic equation. ☐
A11	Given the nature of the roots of a quadratic equation, use the **discriminant** to find an unknown (non-linear) or a **condition on coefficients**. ☐
A12	Determine the equation of a polynominal from a graph (**using roots**). ☐
A13	Use the **remainder theorem** for values, factors, roots. ☐
A14	**Factorise cubic** and **quartic** polynomial expressions. ☐
A15	**Solve cubic** and **quartic** polynomial equations. ☐

A16	Find the **intersection of a line and a curve** or **two curves**.	☐
A17	Find if a line is a **tangent to a polynomial**.	☐
A18	Use the **laws of logarithms and exponents** to simplify and evaluate expressions.	☐
A19	Sketch a **logarithmic** or **exponential function** or their **inverse**.	☐
A20	Solve equations involving **logarithms**.	☐
A21	Solve equations involving **exponentials**.	☐
A22	Given two pairs of corresponding values of x and y, **solve for a and b equations of the following forms $y = ax^b$ or $y = ab^x$**.	☐
A23	Use a straight-line graph to **confirm relationships of the form $y = ax^b$ or $y = ab^x$**.	☐
A24	**Use relationships of the form $y = ax^b$ or $y = ab^x$**.	☐
A25	**Develop mathematical models** for situations involving **logarithmic** or **exponential functions**.	☐
A26	Use the notation $\boldsymbol{u}_n$ for the nth term.	☐
A27	**Evaluate successive terms** of a sequence from a **recurrence relation**.	☐
A28	**Determine a recurrence relation** from given information.	☐
A29	Decide when a **sequence** has a **limit**.	☐
A30	**Evaluate** and **interpret** the **limit of a sequence**, where it exists.	☐
A31	**Apply A1–A30** to problems in familiar or unfamiliar contexts.	☐

CALCULUS

	Calculus
C1	**Differentiate an algebraic expression** which is, or can be simplified to, an expression in powers of x (or another variable), such as differentials of px^n and sums and differences.
C2	Differentiate **negative** and **fractional** powers.
C3	**Find the gradient** at a point on curve and vice versa.
C4	Determine the equation of a **tangent** to a curve at a given point by **differentiation**.
C5	Determine the **equation of a curve** given the equation of a tangent and a point on the curve.
C6	Determine where a function is **strictly increasing or decreasing**.
C7	Find **stationary points**/values.
C8	Determine the **nature** of stationary points.
C9	Determine the **greatest** and **least** values in a **closed interval**.
C10	Sketch the graph of an algebraic function by **determining stationary points** and their nature as well as **intersections with the axes** and behaviour of $f(x)$ for **large positive** and **negative** values of x.
C11	**Differentiate trigonometric functions** of the forms $p\sin(ax + b)$ $p\cos(ax + b)$ $(\cos x)^2$
C12	**Differentiate a composite function** using the **chain rule**, such as $y = \sin^2(3x)$ $f(x) = (4x^3 + 8x)^{-\frac{1}{2}}$

C13	**Integrate an algebraic expression** which is, or can be simplified to, an expression in powers of x (or another variable) such as integrals of px^n and sums and differences.
C14	Integrate with **negative and fractional powers**.
C15	**Evaluate definite integrals** of functions with **limits** which are integers, surds or fractions.
C16	**Find the area** between curve and a line.
C17	Find the area **between two curves**.
C18	**Integrate polynomial functions** $(ax + b)^n$.
C19	**Integrate trigonometric functions** $p\sin(ax + b)$ $p\cos(ax + b)$
C20	Solve problems using **rate of change**: determine and use a function from a given rate of change and initial conditions.
C21	**Solve differential equations** of the form $\frac{dy}{dx} = f(x)$.
C22	**Apply C1–C21** to problems in familiar or unfamiliar contexts [e.g. Optimise, great/least, complex area] including determining a mathematical model.

GEOMETRY

Geometry	
G1	**Find the distance between points** (using the distance formula). ☐
G2	Find **gradient** from **two points**, an **angle**, the **equation of the line**. Use $m = \tan\theta$ to calculate a gradient or angle. ☐
G3	Find equation of a line **parallel** or **perpendicular** to a given line. ☐
G4	Determine whether or not two lines are **perpendicular**. ☐
G5	Calculate the **midpoint**. ☐
G6	Find and/or use properties of **medians**, **altitudes** and **perpendicular bisectors** in problems involving the equation of a line and intersection of lines. ☐
G7	Find the **centre** and/or **radius** of a **circle** from its equation/other data. ☐
G8	Determine and use the **equation of a circle**. ☐
G9	Find **equation of a tangent** to a circle. ☐
G10	Determine the **intersection of line and circle**. ☐
G11	**Determine** if/when a **line** is a **tangent** to circle. ☐
G12	Use properties of **tangency** in the solution of a problem. ☐
G13	Determine the relative positions of circles. ☐
G14	**Vector algebra** (such as addition of vectors, finding vector **b** given vectors **a** and $\overrightarrow{AB}$). ☐
G15	Use and find **unit vectors**. ☐

G16	Use vector algebra properties: if **u**, **v** are parallel then **v** = k**u**.	☐
G17	Simplify **vector pathways**: determine the resultant of **vector pathways in 3D**.	☐
G18	Find if three points in space are **collinear** and work with **collinearity**.	☐
G19	**Find ratios** when one point divides two others on a straight line.	☐
G20	Given a ratio, **find or interpret a 3rd point or vector** e.g. find the cooordinates of a point dividing AB in the ratio 2:3.	☐
G21	**Evaluate a scalar product** given suitable information and determine the angle between two vectors.	☐
G22	**Apply properties of the scalar product** e.g. if **u**, **v** are perpendicular then **v.u** = 0	☐
G23	Use the **distributive law a.(b + c) = a.b + a.c**	☐
G24	Use the **unit vectors i, j, k** as a basis for 3D space.	☐
G25	**Apply G1–G24** to problems in familiar or unfamiliar contexts	☐